FENCING

or

THE SCIENCE OF ARMS

By

SALVATOR FABRIS.

TRANSLATED

by

A. F. JOHNSON.

This book contains a representation of an early 20th century typescript by A. F. Johnson, currently owned by the Howard de Walden Library and housed in the Wallace Collection, containing his unpublished translation of the 1606 treatise of Salvator Fabris. Since it is not possible to scan the book, it was photographed by Guy Windsor and these photos were carefully transcribed by Michael Chidester, and then formatted to resemble the original as closely as modern technology allows. No attempt has been made to correct spelling, grammar, or formatting errors present in the original.

Published by HEMA Bookshelf, LLC
411a Highland Ave #141
Somerville, MA, 02144
www.hemabookshelf.com

ISBN 978-1-953683-21-2 (hardcover, unillustrated)
ISBN 978-1-953683-22-9 (softcover, unillustrated)
ISBN 978-1-953683-23-6 (hardcover, illustrated)
ISBN 978-1-953683-24-3 (softcover, illustrated)

Typeset in LTC Remington Typewriter.

To His Serene Majesty, the most Powerful Christian IV., King of Denmark, Norway, Gothland and Vandalia, Duke of Schleswig Holstein, Stormarn and Ditmarsch, Count of Oldenburg and Delmenhorst, &c.

I am confident that all who read this work of mine will recognise that the many benefits received from your Serene Highness are the cause, which has urged and impelled me to publish to the world these my labours. I have wished also to help professors of the science of arms by showing them those instructions and rules, which after long use and continual practice and from observing the errors of others I have found to be good. I hope then that a work based on such principles will find merit, especially as it is under the protection of your Serene Highness - a work as worthy by reason of the excellence of its subject as it is glorious through the approval of your high judgment. To you, therefore, my benefactor, my king and a prince of incomparable valour as much in civil government as in the practice of arms, a true hero of our times, I have dared to dedicate my work; for since its inception is due to you, I am bringing it forth to the sight of men under the same protection. I know moreover how useful to the world and necessary to good men this art is, bringing honour to anyone who practises it aright either in the defence of his prince, his country, the laws, his life or his honour. Will your Serene Majesty therefore deign to

receive into your favour not only the work, but the devotion
with which, your humble and obedient servant, dedicate it.
Meantime I will pray the Divine grace that long life may be
granted you for the well-being of your blessed subjects and
the good of the world, and that by grace you may obtain sal-
vation in the world to come.

Your Serene Majesty's most humble and

devoted Servant,

SALVATORE FABRIS.

TO THE READER.

Marvel not, Reader, if you see a man of the sword, unaccustomed to the schools or the circles of literary men, presuming to write and print books; rather rejoice at seeing the science of arms and the knowledge of the sword reduced to rules and precepts, and like the other arts to a teachable form, wherein the curious and eager men of arms may learn by turning the leaves.   More than others should men of arms rejoice, in that men of learning and science have never translated their arts from theory to practice, as now a man of arms has brought his from practice to true theory.   To him is owed much greater faith, because he has had a thousand experiences in his own case and in that of others of what he has written.   Here then, Reader, is my book on the science of arms, illustrated with plates suitable to each case; to these plates and dumb images, as it were, our words give life; the plates will demonstrate and our words will interpret the effects and principles treated of in the book.   We have written in our mother tongue, Italian, dispensing with flowers of rhetoric and elegance of style, not thinking shame to acknowledge our little learning, or, following the example of a very famous captain of our age, to declare that in our youth we could not wield both the sword and the pen.   We believe however that we have dealt adequately with what is required in this art and have tried as far as in us lay to avoid obscurity and prolixity, although in

so subtle a subject it is difficult to preserve the necessary
brevity. We have shunned the use of geometrical terms, although
swordsmanship has its foundations more in geometry than in any
other science. Simply and as naturally as possible we have
tried to bring the art within the capacity of all. For what we
have written and demonstrated we require no praise or reward,
for it was never our intention to publish it to the world; but
if in it there is anything worthy of merit, it should be as-
cribed to his Serene Majesty our King, through whom we have written this work, and at whose command this book is brought to the
light of day. We will not speak of the nobility and excellence
of this profession, for it is in itself so glorious and resplen-
dent that it has no need of our words, nor is there any man so
ignorant as not to know, that by its kingdoms are defended, re-
ligion spread abroad,injustice avenged, peace and the prosperity
of nations established. We wish only to say that after
acquiring
this inestimable knowledge a man should not become puffed up nor
use it violently to the detriment of others, but always with
moderation and justice in all cases, thinking that the last victory of all rests not in his own hand, but in the just will of
God; and may He grant us abundance of his saving grace.

GENERAL DISCOURSE OF THE FIRST BOOK.

THE PRINCIPLES OF THE SWORD ALONE.

In opening our promised work we shall begin with the
sword alone, for on the knowledge of the sword depend the
principles of all other arms.   Many rules will be given
which may serve excellently for the sword accompanied by
the dagger or any other arm. He who can use the sword
alone well will easily learn to use it in conjunction with
other arms. You must know then that the rules of the sword
are founded on four guards in which are formed all the positions and counter positions. From them arise the _times,_
counter-times, disengagements, counter-disengagements,double
disengagements, half disengagements, and re-engagements; nor
in short can anything be done in attack or defence which does
not partake of the nature of one of these four guards. They
are differently formed, as will be seen in the accompanying
plates. These we have introduced in order that you may recognise with what variations of position of the sword, feet
and body, they are made. We shall describe the nature of
each guard in its place and the plates will show the results
which may arise from them. The discourses will be such that
you will easily see when to apply the various rules, and how

to the best advantage you must approach your adversary in
order to come within presence. Though one who understands
the art may approach as he pleases, since in whatever position he is he will succeed by his knowledge of distances,
weak and strong positions, exposed and unexposed parts.
Nevertheless it is certain that one position is better than
another, and a man may approach with more security when he
parries his arms in the proper manner. When within distance
he must proceed in various ways, according to the changes
made and the opportunities offered by his adversary, and according to the distance in which he finds himself.   The
distances are two, and what is good in the one is not so in
the other. These distances control the whole attack and defence, as we shall explain. First we shall describe the four
principal guards, why they are called, _prime, seconde, tierce_
and _quarte,_ and the origin of these names. Then we shall
treat of the divisions of the sword, then of counterpositions,
distances, and some other matters which we consider necessary
and useful to the good student of this art.

DESCRIPTION OF THE FOUR PRINCIPAL GUARDS
AND THE ORIGIN OF THEIR NAMES.

The four guards arise from the four faces of the hand and the sword, that is to say of the two edges and the two surfaces; and these produce four different positions. _Prime_ is that position which the hand takes in drawing the sword from the scabbard, when the point is turned towards the adversary - all the guards especially with the sword alone must be formed with the point so directed. When the hand is turned slightly upward we have _seconde_, and _tierce_ when the hand is in its natural position turned neither up nor down. When the inside of the hand is turned upwards we have _quarte_. The hand in turning can take these four positions only, and being in _prime_ cannot go to _quarte_ without passing through _seconde_ and _tierce_; so the name _quarte_ is given to the last position. _Prime_ is the most suitable position for grasping the sword, although it can be done in _seconde_ or _tierce_: but with the hand in _quarte_ the sword cannot be drawn from the scabbard. You must know that nothing can be done which does not arise from one of these four positions approximately; we say approximately, because, if you consider, you will find that there is a great distance between one guard and another ow-

ing to the width of the surface of the sword and of the hand,
so that between <u>prime</u> and <u>seconde</u> there is a mean, where the
hand might stop, and similarly between <u>seconde</u> and <u>tierce,</u>
and between <u>tierce</u> and <u>quarte.</u> Therefore one might say that
there were four legitimate guards and three bastard, since
each bastard resembles the two, between which it is formed.
But to avoid the confusion of so many terms we shall speak
only of the four legitimate guards, which will serve very
well for the three bastards also; for the quality of the
guard is considered not only from the position of the hand,
but also from the direction of the point, wherein lies the
force of the guard. Therefore we shall divide the guards
into these four only, especially as with the sword there are
only four methods of hitting, that is on the inside, on the
outside, below and above. The great differences between one
guard and another will be explained when we treat of their
natures, when we shall consider the various methods of defence, and the changes made in hitting, according to whether
they are formed with the sword extended or withdrawn, high
or low; we shall then treat of the nature of each one separately.

THE DIVISIONS OF THE SWORD: THE FAIBLE AND THE FORTE.

The blade of the sword is divided into four parts; the first is the part nearest the hand, the second quarter extends to the middle of the blade, the other two extend to the point. The first part near the hand is the strongest for parrying, and there is no thrust or cut, delivered by the strongest arm, which, if parried in this part, the sword cannot defend and resist without disorder, if the rule and the time is observed, as will be explained. The second part is somewhat weaker, but still it will defend well enough, if you parry where your adversary's sword has less strength. The third part is not good, especially against cuts, unless strengthened with the adversary's body at the moment of parrying; this will be explained in considering the defence. The fourth part is entirely bad and must not be thought of in the defence, although in the attack it is the strongest and most deadly. It is also true that a cut made with half the third part and half the fourth is a strong attacking stroke, whereas a cut made with the third part only would not be one half so effective as a cut made with the fourth part as well. The first and second parts then are only to be used in the defence, and the third and fourth in the attack; so that the sword is divided, one half into the defensive part, and the other the offensive.

METHOD OF FORMING THE COUNTER-POSITIONS, showing

THE POSITION OF THE ARMS AND THE BODY, AND WHEN

THEY ARE TO BE FORMED.

If you wish to form a sound counter-position, the position of the body and arms must be such that without touching
the adversary's sword you are defended in the straight line
from the point of his sword to your body, so that without making any movement of the body or the sword you are sure that
your adversary cannot hit you in that line, but that if he
wishes to attack he must move his sword elsewhere, with the
result that his _time_ is so long, that there is every opportunity to parry. But in forming this position care must be
taken that your sword is held in such a way as to be stronger
than your adversary's, so that it may offer resistance in defence. This rule can be observed against all positions and
changes of your adversary, whether accompanied by the dagger
or any other defensive weapon, or when you use the sword
alone. He who can most subtly maintain this guard will have
a great advantage over his adversary.

But it often happens that when you form this guard, your
adversary forms another against it. Often also this guard is
formed so far out of distance that your adversary can wait
until you begin to move your foot against him, and at the moment of year advance change his line, so that you are disconcerted by another counter-position. Therefore you must be

full of devices and be able in a moment to take up another
position of advantage against that of your adversary and
make a fresh guard, unless you are so far within distance
that you can hit him daring this change, and if in changing
he has not retired, since if he had retired you could not hit
him even if you had been within distance. You must then take
up another counter-position and approach at the same time, to
regain the same distance as before. In forming this counter-
position you must bear in mind the rule, that the body must
be so far distant that the adversary cannot hit, or, if you
have approached within distance so that he could hit by advancing his foot, you must form the counter-position without
moving the feet. In this way, if the adversary should attempt
to hit during the movement, you could parry and hit him, or
break ground; in the latter case his sword, would not reach.
But if in moving your weapons to take up this advantage, you
have moved slowly, you could then abandon your object and hit
at the very moment in which your adversary advanced to attack,
parrying at the same time. So that if the first movement is
made without violence, you can abandon your attempt and make
another, as opportunity offers. In short, if you wish to
get within distance with some safety, you must first form the
counter-position, and if disconcerted by your adversary's
counter-position, it will be better to break ground than to
approach, until there is an opportunity to get an advantage.

EXPLANATION OF THE TWO DISTANCES, WIDE AND CLOSE, AND
HOW TO ACQUIRE THE ONE OR THE OTHER WITH LEAST DANGER.

You are within wide distance when by advancing the rear
foot to the front you can make a hit.   After forming the
counter-position a little out of distance, you must begin to
advance the foot in order to get within the required distance.
But you must be on your guard, lest your adversary, being
steady, at the moment when you move your foot to advance it,
should advance his too and hit at the same time.   Therefore,
you must move it very carefully, remembering that your adversary may effect something during the movement.   After forming
the counter-position you must endeavour to throw him into disorder, or at least make some feint in order to have an opportunity to hit.   Thus prepared for what may happen you are more
guarded and can better resist attack.   When you are within
wide distance and your adversary makes some movement of his
foot, provided he does not break ground, you can hit him in
the nearest exposed part, even if he has not moved his weapons.
This could not be done if he moved his weapons and stood firm
on his feet, the reason being that a movement of the foot is
slower than that of the weapons, and therefore he could parry
before your sword arrived, while he remained steady; if there
were no other way he could protect himself by breaking ground,

so that your sword could not reach. Being thrown into disorder you would then be in danger of being hit before you had
recovered. Therefore whenever he gives an opportunity without
moving his feet, it would be better to approach within close
distance in that time. In that distance you can reach with
the sword by merely bending the body, without moving the feet,
and the adversary is forced to retire to get out of such danger. If he does not move you could hit him even though he
retained the advantage of the counter-position. If your adversary does not move, you can sometimes make a hit by judging
the distance from the point of your sword to your adversary's
body and the distance from the _forte_ of his sword. If you consider both how much you must advance the point and how far you
must move it from the adversary's _forte_, and understand that
the time required for him to parry is the same as for you to
hit, the sword will arrive before he has parried by the advantage of having moved first. If you see that his body is
little exposed, as may happen, since one guard covers more
than another, you can then attempt to hit in the exposed part,
and as he moves to the defence change your line and hit in
the second exposed part.

These rules apply within close distance. If you are within wide distance and wish to advance within close distance,
the danger is greater when the adversary stands steady on his
guard, because if you raise your foot to advance it, you give
him an opportunity to hit and retire, so that at the end of

the movement you would be at the same distance, that is wide
distance, and would have obtained nothing. All this is due
to the fact that you cannot move your foot in less than two
times, the one in lifting it and the other in putting it to
the ground. For this reason some push the foot forward by
scraping it along the ground, which is well enough in the
hall, but in the street is likely to lead to a fall because
of the many unevennesses. It is better then to lift it to
make sure of not stumbling. Therefore in carrying the foot
within close distance you must first form a good counter-position, and then lean all the weight of the body on the rear
foot as you lift the forward foot, so that if in that moment
your adversary should thrust you would be able to parry and
to hit by bringing your foot to the ground, or even extend
that movement which you had already begun beyond your first
design, in order to reach more certainly in case your adversary broke ground in making his hit. If the adversary has not
moved the pupil must after raising the foot carry it within
close distance in such a way that the weight of the body rests
on the rear foot, and is no nearer than when within wide distance. After putting your foot to the ground you could then
by merely bending the body hit on the slightest movement if
the adversary in the line exposed nearest to your point.

If you did not wish to wait you could hit in the manner
already described. If while you are carrying your foot within

close distance, your adversary should retire, you would remain within wide distance, and must bring the weight of the
body from the rear foot to the forward and then bring up the
rear foot close to the other. In approaching within close
distance always take care that the body does not approach
with the foot, but remains in the same position as before,
and after bringing your foot to the ground carry forward the
body. This rule should be observed in every case of requiring close distance, but after hitting you must in recovering
your weapons draw back the body as far as possible and draw
back the foot in such a way that if your adversary follows
you are ready to parry and hit. If you find that your adversary is always breaking ground you must not grow angry and
pursue him. Rather you must then proceed more carefully, for
many feign a retirement with the object of drawing on their
adversary and seeking an opportunity to hit in the moment of
his pursuing. If you follow our method you will avoid this
danger. It is better not to pursue one who flees, but rather
to feign reluctance in order to reassure him and so draw him
on, and then to seize an opportunity which he will not have
time to avoid.

DISCOURSE ON RUSHING IN WITH THE SWORD EXTENDED
AND THE PRINCIPLES OF THE TWO TIMES, SHOWING
THAT IT IS BETTER TO CONTROL THE SWORD AND
OBSERVE THE CORRECT TIME.

There are some who, in endeavouring to hit with the
point, hurl the arm violently forward so as to give it greater force. This method is not good for the reasons which we
shall bring forward. In the first place when you rush in
with the sword, should your adversary anticipate you and defend the part where you intended to hit, you cannot change
your line, as would be necessary, so that the adversary is
sure of his defence. If he has also realised the weakest
part of your thrust and pushed your sword in the direction
in which it is being naturally carried, he will drive it
out of line all the more quickly. His defence will be very
simple without using any force, because if he pushes the
sword in the direction in which it is naturally falling, it
will fall the quicker without any resistance. In this manner
his _faible_ is stronger than the _forte_ of the hitter. Moreover in completing the rush the point of the sword drops so
that it cannot hit exactly the point aimed at, and/at the end
of the extension it is impossible to prevent the arm and sword
from dropping to the great advantage of your adversary. Further after one rush it is impossible to make another without

withdrawing the arm again, which takes so long that, if the adversary has not hit at the first fall of your sword he could hit while you are withdrawing your arm, and recover before the second rush, with excellent opportunity of parrying and hitting even if he did it in two _times_, that is parrying first and then hitting. The rule of the two _times_ then would be good enough against such a method, and all the more successful as those who rush cannot make any good feint; for in feinting they move the foot or the body without advancing the sword, or if they advance it often withdraw it even further than before in order to hit with greater force, a very slow and dangerous _time._

In treating of the rule of the two _time_, we say that, although it may succeed against some, it is not to be compared with the rule of parrying and hitting at the same time, because the true and safe method is to meet the body as it advances, before it has had time to withdraw and recover. If you then pursue you give an opportunity for parrying and hitting again. It has been our experience, that most of those who observe this rule of two _times,_ if they can engage the adversary's sword, generally beat it in order then to proceed with the stroke. This would be successful but for the danger of being deceived. He whose sword has been beaten on the _faible_ certainly cannot hit at the same time, as he is thrown into disorder by the beat. But if he happens to disengage he

causes the sword which has beaten and missed to drop still further, and has an excellent chance of hitting. Even if he made a feint of beating, so that when the adversary disengaged he might beat in another part, he would still be in danger of being hit, because the adversary might make a feint of disengaging and return, and in this way the one who had meant to beat would not be able to parry. Finally it may be taken as established that it is impossible to beat your adversary's sword without putting your own out of line. Moreover sometimes if you attempt to beat the _faible_, according to rule, you meet the adversary's _forte_, which he has p shed forward, so that the beat fails and your adversary proceeds to hit without your being able to prevent him. In dealing with one who does not rush, but controls his sword, even though you beat his _faible_, his _forte_ does not move, so that he can parry. Therefore, we conclude for these reasons and for many others which might be adduced, that it is better to parry and hit at the same time, though with the sword alone great judgment is required to effect the two at one moment. As to controlling the sword or thrusting with violence, controlling it is beyond comparison better, first because he who controls his sword, when it is beaten by the adversary, who means to hit in another line, can let it yield in the direction of the beat, and the _forte_ will still defend, if the sword is held well advanced. Further it is certain that when your sword is beaten, it is im-

mediately freed. Similarly it is more useful to know how to be master of your sword, to engage the adversary's faible and make a hit as opportunity offers, always holding his sword in subjection. If he cannot free his sword he cannot hit. Therefore this rule can be followed only by one who moves his sword without violence, works in such a way as to be always master of it, and if he is prevented by his adversary in any plan can abandon it and adopt another. He will hit at the very moment when his adversary has meant to prevent him, and without deviating his point or withdrawing it he will be able to carry it on to the adversary's body. The principle to be observed is this, that in proceeding to make a hit either by a feint of disengaging or any other change, when once you have begun to approach, the point toward the adversary, you must continue until you reach the body; for if you check the sword in order to disengage or change your line you will not arrive in time. This principle cannot be observed by one who rushes, so that the difference is easily understood. Moreover the sword which is held firm and accompanied by the foot and the body has greater force and exactness. He who so hold it always controls it and does not let it drop after a hit. He has only to withdraw his foot in order to bring his body to safety, unless he has passed, and to engage the adversary's sword again. If your adversary as you withdraw, pursues or advances, you can hit again, defending at the same

time. All this is because of the union between the sword, the feet and the body. If this rule is observed in the manner we have described, your parrying will be safe, whereas with the rule of the two _times_ it is false; this will be better understood in its place.

DISCOURSE ON CUTTING. How many cuts there are and how they are made, their nature, and whether it is better to use the point or the edge.

The principal cuts are four; they are delivered in different ways and in different directions, as will be seen in the plate which follows (pl.1.) with their names. The names are derived from the four principal cuts, that is _mandiritto_, _riverso_, _sottomano_ and _montante_. They are delivered in various ways, for some deliver them from the shoulder, some from the elbow, some from the wrist, and some again from the shoulder but with the arm extended and stiff, keeping the point always directed towards the adversary. In making the first cut from the shoulder, the arm is raised and makes a circle with the sword in order to strike with greater force. This is the worst of all because of its excessive slowness and because you may easily be hit as you raise the arm, as you let it fall, or after it has fallen; for as the sword is not supported by the adversary's weapon or body, there is nothing to prevent it from passing on behind his back; or if

the hit is made downwards, the sword is in danger of being broken on the ground. In either case so much time is lost that your adversary may easily hit. The second method from the elbow also carries the hand out of line, both when it is raised and when it falls after missing, so that in this case too you may be hit, but not so easily, as the sword does not make such a large circle, nor does the raising of the arm un-cover so much, nor the sword fall so far. Therefore as the movement is quicker and you remain better covered, this me-thod is better than the first. The third method, made from the wrist downwards with the arm straight, although the sword makes a circle is beyond comparison better than the two first described, since the body is more covered. Nor can you be so easily hit, since it is quicker, and the point in falling re-mains in such a position that you can parry with the _forte_ either thrust or cut, and can cut again. Similarly the fourth method with the arm stiff and extended is far better than the two first, since you hit without making a circle with the sword, raising it little or nothing. The sword is allowed to fall on an exposed point, and when your adversary makes a circle with his sword in order to hit, with this fourth me-thod you can continue your stroke, as you will certainly hit before his falls. You will be all the more secure if you have worked with the feet and the body, as you should, be-cause if you remained upright when your sword fell, you could

not recover in time, especially if your adversary's cut had
been made from the elbow. But if you lower your body the
sword is more quickly recovered and has less distance to move
in returning to the defence, for as you hit with the arm stiff
and extended without bending the wrist, the sword still remains in front and can easily return to the straight line.
For this reason the fourth method is better than the two first
and in defence better than the third, although it appears to
us that the third is much freer or less restricted, and without requiring so much strength has more variety and can more
easily deceive the adversary.

He who wishes to make a cut with safety, must wait a
fitting opportunity, since he cannot make the stroke in a moment, and the time might have passed before the sword arrived. You can make a feint in order to put the adversary in
subjection, and whilst he is parrying the cut, thrust at him,
or make a feint of a thrust, and cut. The latter method
would be necessary if you wish to move without waiting, for,
if your adversary remained steady, it would not be good to
make a feint of a cut in order to thrust, owing to the length
of the movement, during which you might be hit. You can, as
has been said, make a feint of a thrust in order to cut, and
even if he parries the cut, still make a thrust. Further the
feint of a cut, when your adversary stands firm, is not good
because of the two times involved in raising and dropping the

arm. All the cuts are very long, and he who cuts cannot do so in the _time_ of a parry (we speak of the sword alone), whilst the adversary has always time to protect himself and even to make a hit when you are trying to parry. It is true that in parrying you can put your adversary in subjection and deprive him of the power to do anything, and even hit him before he can save himself; but of this we shall speak when we treat of the defence and the attack.

Since cutting is not very useful we shall not dilate on it more than is necessary to show the respective merits of the thrust and the cut. Still it is well to be acquainted with both. In cutting greater strength is required, which is a disadvantage. The sword, if it misses, is thrown into disorder, and the body too. Recovery is not so easy, so that you are in more danger than with the thrust. Further it is less deadly; so that in all respects thrusting is more advantageous. With the point you reach further, more quickly and can more easily recover. In brief thrusting is more noble and excellent, for it includes all the subtlety of arms, whereas in cutting there is neither the _counter-time_ nor the _time_, since for the most part two _times_ are involved. We do not intend to discuss this further than we have done in the preceding chapter in relation to the two _times_, but to consider the more subtle, difficult and profitable points. If for example two men were opposed, one who excelled at the cut

and the other at the thrust, without a doubt the latter would prevail for the reasons we have given, though his opponent were the stronger man. We conclude that it is better to use the point only, especially in engagements corps à corps without armour. With armour we should deem it good to use both; so too against a number of opponents, for the cut causes greater confusion and may parry several thrusts.

In good and false Parrying, and on some who, with the sword alone, parry with the left arm.

The parry partakes of the nature of fear, for he who did not fear some disadvantage, would not put himself on the defence which we may call obedience and subjection, all the more so when it is forced. He who does not wish to be hit is forced to parry. When we can put our adversary under this obligation, we consider it a great advantage, for while he is under the necessity of parrying he may be hit in the line he uncovers by his movement, so that his defence proves vain. Therefore some say that parrying is false, which we admit, when it is done alone, for when you make a feint in one part and hit in another, your adversary moves to the parry, and, thinking to defend himself, is deceived by the feint, when he might, instead of parrying, have allowed the thrust to pass. To avoid thrusts is always better - that is with the sword alone, for with the sword and the dagger you can parry with one weapon and hit at the same time with the other, so that

the defence is easier. But with the sword alone you must be more judicious, as it has to perform the two functions of defence and attack at the same time, in order that the parry may be safe. If you are forced to parry by a cut you must give the opposition with your _forte_ where the adversary's sword is about to fall, and at the same time drive the point in with great swiftness, in order that it may arrive before your adversary's falls, so that he may neither avoid it nor be able to hit. This is an excellent rule, because the cut is shorter than the thrust. If you see that you cannot arrive with a _time_ thrust, there is no need to parry, for it follows that your adversary cannot reach. If you are in doubt, you can withdraw the body a little and let his sword fall, and hit at the end of its fall. If you wish to parry knowing that you cannot hit, you must still carry your point as if you meant to hit, as this prevents the adversary from changing his line. Thus you free yourself from subjection and force your adversary to take the defensive, since he is threatened with a _time_ thrust, and his subjection gives an opportunity for a hit. Hence it is never necessary to parry without hitting, or making a feint of hitting in order to force your adversary to parry, so that you free yourself from danger and at the same time place him in danger. It often happens that one who attempts a cut makes a large circle, so that you may hit him and recover before his sword falls. For

in addition to the fact that the cut is slower, as we have said, it is also shorter. This you may do by understanding your adversary's movements and his distances. When the distance is so great that you cannot reach, you must make a feint of hitting while the adversary is making his circle in order to make it fall all the more precipitately and then sieze the opportunity to hit in the part uncovered by its fall. This is instead of parrying at a great distance. But within close distance you may hit before your adversary's sword descends,since the thrust is finished before the cut,so that by withdrawing the left foot and recovering the body you may get into safety and your adversary fail to reach. It is true that this stroke would not be so deadly, since with the parry you can advance further and hit with more vigour and can pass right to the adversary's body. If you do not desire to pass it is necessary to understand how to maintain close distance and to control the feet in such a manner as to break ground in order to avoid a hit. This method is very successful owing to the slowness of the cut and because your thrust reaches further and you move with greater swiftness, so that you can always recover. If these rules are observed he who attempts to cut will always be hit, as we have already said. If in this place we must mention those who first cut at their adversary's sword in order to throw it into disorder, and then hit, we shall not treat of them at greater length, because he

who understands <u>time</u> and disengaging can easily save his sword against a beat.

In defence of the thrust you must understand that its effect is swifter and more deadily. In defending against it more subtlely and ingenuity are required but less strength. To parry is more dangerous and deceptive owing to the rapid changes which it renders possible. It often happens that although you use the subtle combination of the parry and thrust at the same time, you are still deceived because your adversary, seeing your plan, removes his body out of the line, allows the point to pass and then hits in the part uncovered by the movement; so that avoiding is more subtle in defence and attack against one who makes a <u>time</u>-thrust, if its use is well understood. You must then understand both this and the parry and know how to avail yourself of the one or other as occasion offers. It is even more effective to use both methods together, making half a motion to defend with the sword and half a motion with the body. This defence is quicker, disorders the sword less, and deprives the adversary of the advantage of changing his line. Such avoiding is more useful with the sword alone than with the sword and dagger, but the defence partly with the body and partly with the weapons may be observed in all cases.

As there are many, who, although they have the sword alone in their hands, base their defence rather on the bare

hand than on the sword, we must say something of these. We
say then that such a method of fencing should rather be called the sword and glove, than the sword alone, because they
not only parry with the hand, but seize the weapon and hold
it. We do not think this would succeed with the naked edge
of a sword. To protect yourself with a naked hand is a mis-
erable defence. Still we will show how to proceed against
such men, and how best to use the method in order to save the
body and the hand itself, and how to deceive the adversary.
It is true that those who use the hand in this way can make
larger movements with the sword, since the hand defends in
any case of the adversary's making a hit, and successfully
when he makes his hit simply in the straight line without
disengagement or feint. Still it will not succeed if you
hold your sword pointing slightly upwards, just so far that
you are certain that your adversary cannot pass or hit before
you have directed your point against him. In this position
the adversary can neither engage nor reach your sword when on
guard; you must take care to attack in an oblique line when
you hit, for this is very deceptive to one parrying with the
hand, since in the very act of hitting the sword swerves
aside. After engaging the adversary's sword, being within
distance, taking the _time_ and the opening, you may proceed to
hit by making sure that your point reaches the part aimed at
as soon as it takes that direction. You will certainly make

a hit before your adversary can find the sword with his hand, if he has not had time to break ground. Moreover you can use various feints according to the position in which you find his hand. Finally it is often easier to hit those who use the hand rather than defend with the sword, because they trust to the hand and take no account of the _forte_ of their own sword, merely endeavouring to prevent it being engaged by the adversary; therefore they keep the _forte_ withdrawn, so that they are more exposed. Therefore it is easier to hit them and recover before they have brought the _forte_ forward to the proper distance.

It is still easier to succeed against those who first parry with the hand and then rush, which is done by most of those who follow this principle of defence. Nevertheless he who is well trained although he makes use of his hand, controls his sword and observes the _time._ It is well to know how to do this in case of need, but not as a fundamental principle, as we have said above. He who understands what can be done with the hand, knows the converse even better. As a general rule the hand should never be used except when you can reach the hilt and come to grips, a matter which does not concern us, who wish to treat only of the defence, the methods of making a hit and the advantage of arms, and not of scuffling. However this arises sometimes from accident, so that at the end of the book we shall say something of it.

26

But when such a point is reached, the greatest danger is already passed, of which it is more necessary to treat, in order to show how to avoid it with safety and with damage to the adversary.

Finally to show the true method of using the left hand we say that, when your adversary is about to hit, you must parry with the sword and hit, but it is good at the time of doing this to carry the hand to that part, where his sword might hit, so that the hand will defend the body and exclude the adversary's sword without a beat. This is a good principle on every occasion when there is time. This is a better method, because,the hand is not in such great danger and the body is better defended, nor can the adversary so easily be aware of it, as his sword is not molested. If he proceeds to hit he finds the path closed; if he refrains, you can hit him without being disordered. This then is the safest method of using the hand. He who considers its principles will find in it great advantages and subtleties of defence. Many things we omit for the sake of brevity, as our purpose is to deal with fundamental principles only, from which can be deduced countless rules, some better than others; for this subject is so large that it is difficult to find an end.

27

<u>ENGAGING THE SWORD.</u> How it is done and when
completed.

To engage the sword is to gain an advantage over it; it
is a kind of counter-position with some difference, because
often you have engaged, the adversary's sword without completely
closing the line from his point to your body. But it has this
advantage, that your adversary cannot hit without passing your
<u>forte</u>, which is so near his point, that you can find the point
while he is moving to make the lunge. The counter-position
is not considered well formed except when the line from his
point to your body is fully defended. But the same advantage
may be obtained by relative strength, so that you are considered to have engaged, when you are sure that your sword is
stronger than the adversary's and cannot be pushed aside, but
can push his aside. When on guard and wishing to engage the
adversary's sword, you must carry your point towards his,
with the fourth part of the blade against his fourth, but rather more of your fourth part than of his, for that little
more, though little, will be enough to give you the advantage,
when you have engaged his sword at the weaker part. You must
bear in mind that the sword is always stronger in the line in
which the point is directed, and in order to advance in that
line ——————— you must know how to carry your body and
sword in such a way that their strength is in the same direc-

28

This depends to a great extent on the wrist, as will be seen in the plate illustrating the guard on the inside, which is the most difficult. You must also take care, in trying to engage the fourth part, to keep your point at such a distance from the adversary's sword, that he will not have time to push forward the third or even the second part, with the result that while meaning to engage his _faible_ you would have engaged his _forte_. This might happen owing to the distance between the two swords, for the amount you can push your sword forward before engaging is the same as the distance. You must move at the same time as he moves, otherwise you might be hit. Moreover, although the space between the two points were small, while you were advancing to engage, the adversary might perceive this and make an angle, which would strengthen him and bring him away from your advance. If you should push on in order to hit when within distance, his _forte_ would have penetrated so far, that if you had moved in order to engage his sword, you would be unable to defend yourself, and would be hit. Further, if while you are trying to engage, he moved his body away from your point, he could pass right on to your body, before your sword had returned into line. To prevent your adversary's doing this you must first consider the distance between your bodies, and then move forward to engage his sword, carrying your sword without constraint so as to be free to abandon your first plan, when your adversary seizes

the opportunity, and drives the point on to his body, bringing the _forte_ where you intended to put the point. In this
way you will hit the adversary at the moment when he is pushing forward. You must remember that this rule applies to the
guard on the inside, for, if on the outside, you must abandon
your first movement and drop the point under the adversary's
sword to the right side, carrying the _forte_ where you meant
to put the point. On this line too the present method is very
successful if you similarly do not touch the sword in seeking
to engage. The nearer the adversary is, the better and safer
the method is. The advantage is in having brought your _forte_
against his _faible._

It often happens that the adversary, finding his sword
is not molested is not aware that you have already gained the
advantage, whereas if you touch his sword he more easily
realises the fact, and can disengage or retreat or change his
guard, in order to free himself, so that you lose your first
advantage. Moreover if you touch the sword, you impede and
disconcert yourself, so that if a _time_ comes to hit, you cannot take it because of the resistance of your adversary's
sword. Even when there is no resistance and the adversary
disengages, you cannot prevent your point dropping a little,
so that the _time_ is lost. But if you keep your sword suspended, it is the more ready for every opportunity, there is more
use made of _time_,and there is no necessity to force his sword,

which often leads to scuffling. If you do not touch swords,
that cannot happen. When then you advance to engage your adversary's sword and he moves to meet you at the same time,
the one who first yields with the sword and drives on to the
body, can hit before the other touches swords, or in the same
instant. If you do not wish to try a hit, you can lower your
point towards the ground to prevent the adversary's engaging,
and if he follows it you can thrust while his sword is falling. There are many other ways of preventing your adversary
from engaging your sword, except when the point hits, especially if you have won the advantage of the _forte_ against the
faible and the swords are in position. In endeavouring to acquire the advantage over the adversary's sword you must take
care not to advance the point so far in your desire to be the
stronger, that he can pass in one line or another, before you
can direct your point. If you observe these rules you will
without doubt gain the control of your adversary's sword,
which is the first part of victory. Though your adversary
takes advantage of the _time_ he will still be hit. To prevent
your getting this advantage he will have to retreat, changing
the position of his body and sword and to adopt new devices,
which are countless. The one who is more subtle in his movements will maintain his sword the freer.

31

ON TIME AND COUNTER-TIME, which are good and which
false. How to deceive the feint of a _time_, offered
by the adversary so that he may make a _counter-time_.

A movement made by the adversary within distance is called a _time_. For whatever is done out of distance can only be
called either a movement or a change of front. _Time_ then
means an opportunity to hit or win some advantage over the
adversary. This movement is given the name of _time_ among the
movements of fencing in order to convey the idea that at a
given point of time it is the only possible movement. When
the adversary moves, if you perceive an exposed part and are
ready to hit in that part, the adversary will certainly be
hit if within distance. For there cannot be two changes in
one _time_,and therefore you must take care that the _time_ in
which you wish to hit is not longer than the _time_ offered by
the adversary. In such a case he would have a chance to parry before your point arrived, and you would be in danger;
whereas, if you understand the movement, you would succeed.
This is called a _time_-thrust. Besides understanding the
movement you must consider the distance, because if you were
within wide distance, even though your adversary moved his
weapons or body, provided he did not move his foot there
would be no certainty of being able to hit him, even if he
were uncovered,for, if his foot were firm, he could break
ground, so that your sword would not reach, and you would be

in danger. Therefore it would be better to take advantage of his movement to approach within close distance so as to hit with certainty at his first movement with weapons, foot or body, or with both foot and weapons. All these are _times_ favourable for a hit in an uncovered part. The success would be even greater, when the adversary offers the _time_ unawares, provided he is not retreating. To be certain of success you must be in counter-position, since, if your adversary has moved first, it is clear that he will not be able to parry and hit except in two _times_, so that the stroke will be finished before he has parried, and you will be able to break ground before he hits. It is also clear that he will be unable to break ground, as he might have done if he had remained steady. t is sometimes good to beat the adversary's sword within this distance, even if he does not move his foot, for the reason that, if he offers a _time_ unawares, he will not expect it, as he has not realised that he has given an opportunity of being hit, and therefore he has had time neither to parry nor to break ground.

But you must bear in mind that there are some men who cunningly offer a _time_, that you may attempt a hit, and at the same time they parry and hit. This is called a _counter-time_. Whenever you are hit or make a hit at the moment when your adversary is extended to hit, it is called a hit in _counter-time._ Similarly it sometimes happens that both are hit at the same moment; this is because one of them has not

timed the <u>counter-time</u> well, or that in offering the <u>time</u>
he was too close, or that he has made too large a movement.
To avoid the danger of this <u>counter-time,</u> you must realise
before you make your movement, whether it is so great, that
you could approach nearer, and also whether your adversary
has moved with the intention of enticing you to hit. In that
case you should either not proceed, or you should carry your
sword towards the line uncovered by the adversary, and when
he moves to make the <u>counter-time</u>, you should then change
your line to the part uncovered by his movement, avoiding his
point with your body. In this way the deception planned by
him will be turned against himself. In truth this science
of arms is merely the science of deceiving your adversary
with subtlety.

When therefore you are within close distance you can hit
at every movement or change of your adversary, however small,
provided he does not break ground; for if in giving a <u>time</u>
he carries hit foot back he so lengthens the <u>time</u> in which you
may hit, that he has a good chance of parrying and hitting;
for he being the first to move is also the first to finish the
movement. This advantage he would not have if he stood firm,
and tried to break ground, while you were making a hit; for
your point would arrive before he was out of distance, nor
could he parry. Therefore it is not good to be the first to
move when within close distance, except to retreat. You must

also know that within this distance you may often hit without
waiting for a _time_ by the simple advantage of the counter-
position, and by understanding how to move in making a hit
and how your adversary moves in parrying; also owing to the
fact that there are many exposed parts in such a position.
Therefore you must contrive to have your point so near the
adversary's body, that the time required for your hit is less
than the time he needs to defend himself. You must also contrive that your adversary's sword is so far distant from
yours, that it is clear when you advance that he can engage
only with the _forte_ for then your sword cannot be thrust aside
but will continue on its path to complete the stroke.

All these rules may be observed equally with the sword
and dagger, because the weapons are kept more withdrawn and
there are more exposed parts in which a hit may be made, so
that they will be most effective. You may easily understand
then, how dangerous it is to approach within distance unless
your weapons are in conjunction or without some advantage especially within close distance. You have seen too how _times_
and _counter-times_ are made, how they may be deceived, and which
cannot be deceived.

35

WHAT IS "DISENGAGEMENT, COUNTER-DISENGAGEMENT,
double disengagement, half disengagement, re-engagement,
and how and when they should be used.

When your adversary attempts to engage your sword or to
beat it, and you change from one line to another, before he
can beat or engage, you are said to make a disengagement in
time. If, while your adversary is disengaging, you follow
his movement, which he has begun in order to get the superiority, and let your sword go after his, so that you engage him
in the same line as before, that is called a counter-disengagement. If you have disengaged and your adversary has also
disengaged and you then deceive his engagement, that is a
double disengagement. If, without completing the change from
one line to another, you leave your sword under the adversary's,
you make a half disengagement. If you disengage and, when
your adversary moves to engage or to make a hit, you engage
again where you were before, you are said to re-engage - - -.
To make a successful disengagement you must bend forward —
so that, when the disengagement is completed, the lunge is
completed, if you wish to hit, otherwise you will not be in
time. If you follow this principle, your adversary will not
be able to parry, if you have taken the time, though he may
counter-disengage, if that was his intention in seeking to engage. If he had meant simply to get the superiority or to beat

he would certainly be hit. If in seeking to engage your
sword, the adversary remains steady, then you must disengage
in order to free your sword. This gives him an opportunity
for a counter-disengagement, for he has moved at the same
time as you disengaged. Then to protect yourself you must
make a double disengagement and thrust in the same time,in
which he has meant to hit you with a counter-disengagement.
Some remain steady in seeking to engage in order to make the
adversary disengage and so hit him in the straight line, before he has completed the disengagement. In such a case if
the adversary, who has begun to disengage, returns to the
same line as before, carrying his forte to your faible and
thrusting on to the body, he will save himself and certainly
hit at the moment you meant to hit. The half disengagement
is used when you are in doubt that the adversary may pass to
your body, before you have completed, the disengagement,
since your point would be out of presence and could not hit.
Therefore you make a half disengagement to save time, and remain below the adversary's sword in order to hit, removing
your body out of presence, as we shall explain in its place.
Such a half disengagement is not always used in the first
passes, but more often in the second and third movements, as
the distance is shortened. The effects produced by these disengagements will be seen in the plates.

   ON THE FEINT, and why it is so called; in what manner
and at what time it should be used.

When you feign to hit in one line, and while the adversary is defending himself, hit in another, you are said to
make a feint. You must understand, which feints are good,
and which not. For some make their feints with the feet rather than with the sword, beating the ground as hard as possible, so as to frighten the adversary, and while he is intimi-
dated hit him. This method might sometimes succeed in the
hall, particularly if the floor is of wood,so as to re-echo.
This might sometimes cause the adversary to waver. But on
ground which does not ring, it would not have the same effect.
It is of little or no value in either place against those who
understand the art. For if this stamping is done out of distance, there is no need to waver, since his sword cannot reach.
If it is done within distance, it gives you a chance to hit
in the exposed part at that very moment, or to make a feint of
hitting in that part and hit in the other part, which he expo-
ses in his attempt to defend himself; for he can never defend
one part without uncovering another. Thus one who stamps with
his foot will be deceived through not observing that in seeking to provoke his adversary to offer a _time_, he has himself
offered a _time_. The adversary, being steady, can better judge
the movements than the one who has moved. Thus feints more

often succees if made when the adversary is moving than when he is steady.

Others make the faint with the body and the sword, but do not extend the sword much, so that the adversary may not engage it in his parry, and they may then hit him when his weapons have dropped, or when he raises them again violently after failing to engage. This method may succeed with a timid or ill-trained adversary. But as the sword does not come forward, you know that it cannot hit; therefore you should not move, except to attack at the moment of his feint; or you should make a feint of hitting, so that in his doubt that you have taken the time he will rush to the defence, when you will have ample opportunity to hit. This will be a hit with a counter-feint, and he who made the first feint will be deceived.

There are others who in feinting carry the sword forward, and, when the adversary tries to parry, draw it back to return it with a rush. Neither is this method good, but rather worse than the other; for when you should make only one movement with the sword you make three contrary ones, one in carrying it forward, the second in withdrawing it and the third, greater than all, in thrusting with violence. You fall to grasp that your movement is so slow, that, if your adversary moves at the first movement of the feint, he will hit before your sword has completed the withdrawal, and will easily save him-

self before you can proceed to hit.

To make a successful feint you must advance your sword in this way. When your adversary allows your sword to penetrate so far that you know the _forte_ is far enough advanced to resist his weapon, before he is in a position to parry, you must continue the thrust, as he cannot push the sword away. Then you will certainly hit. If, as you make the feint, the adversary moves in time to parry, you must then change your line and thrust the point on to his body in order to arrive, before he has completed the motion of the parry. This is the true method of making a feint. When you make a feint you must remember that your adversary may hit at that moment. For if you were persuaded that he must first parry, you would generally be deceived; but, if you judged that he might attack, you would be more ready on the defence, and if he does not hit, no harm is done, and your movements will be made easier.

You should know too that feints should be made against the nearest exposed parts, for the sword cannot reach the distant ones, nor hit those which are unexposed. You must not fruitlessly put yourself in danger, but if you realise the distance and what part is exposed you will obtain good results. If you work in this manner the feint cannot so easily be recognised by the adversary, so that, if he does not parry, he will be hit; while, if he parries, you can change line

and hit. It is still better to wait for the adversary
to offer some _time_ or uncover some part, since he would be
sure to conclude that you had taken the _time_ from his movement and would rush to the defence more precipitately, and
you could hit him the more easily; nor could he himself hit
in that _time_, so that your security is greater.

If when within distance you uncover some part in order
to give your adversary a chance to hit, you are said to make
an _appel_. You must consider the distance and be careful that
his sword is not so near that it might arrive before you had
finished the movement of the _appel_. You must decide whether
it is better to advance while he is moving, or to retire in
order to have time to parry and hit. Therefore in making the
appel it is not good to move the feet, because you could
bring them neither forward nor backword in time; besides the
danger of being hit through the slowness of the movement. But
the _appel_ can very well be made in withdrawing or approaching
the body according to the nature of the distance, because the
movement of the body is very quick, and if properly made does
not prevent your raising the feet in time. An _appel_ should be
made when you see that your adversary is about to lunge in
order to encourage him to stick to his purpose. Such an _ap-
pel_ is made to deceive him; but if he perceived it he might
deceive you, as we noted in treating of the deception of _time_
and _counter-time_. An _appel_ is simply giving time in order to

invite the adversary to hit, with the object of hitting him.
When your adversary desires to do something, it is better to
encourage his desire rather than prevent it, so that his action will be more hurried. It is much better to know what he
means to do and to let him do it, than to wait for him to do
something unforeseen. It often happens that you are hit without knowing how or why. Therefore you must know your adversary's intentions in order to resist him better. Attack him
in time and protect yourself.

ON LUNGING AND PASSING.

To lunge is to hit by carrying the right foot forward towards the adversary and withdrawing it immediately after hitting, or to hit by a movement of the body, keeping the foot
firm. To pass is to carry both the feet on right to the adversary's body. It is necessary to understand the lunge, as it
is in the most common use, and therefore must be the first
thing to practise, in order that you may learn how to advance
the point accurately and to the full extent. The hand is
fallible and may hit in a spot different from the one intended, according to the amount of the distance. This depends on
the changed position of the wrist as it is extended more or

less, causing the sword to fall short or go too far, in accordance with the angle of its direction. In order to learn
how to drive the sword sufficiently far you must accompany it
by bending the body forward and recovering it quickly after
hitting, in order to save yourself from danger. Practice is
required to learn how to carry yourself, and when you can do
this well you will find it very profitable, for it will make
the body agile, the feet quick, and give you judgment of distances. You will then certainly make a lunge longer than before practice.

To make this kind of hit well you must stand with your
feet not too far apart, so that you can advance further in
hitting, or according to circumstances withdraw by bringing
back the foot, leaning the weight of the body on the foot
which is to remain steady, so that the other may be more
agile and easy to lift, for these reasons it is not good to
be on guard with the left foot forward, because you cannot
make a long lunge without passing; whilst if you tried to pass
with the rear foot and to return you would find the movement
too long; besides you would go too far to be able to return
in time. For these reasons and many others which we omit it
is not good to be on guard with the left foot forward, unless
you are waiting for your adversary to try a hit, so that you
may at that moment withdraw your left foot, parry, and hit
him at the same instant. This method may succeed, because
the body changes its front and withdraws, the right side re-

maining in front for the attack. But if your adversary does
not come on, you should not attack him, since it is better to
have the right side in front; you can hit in shorter time
and save yourself more promptly, as the foot and the body make
smaller movements. After hitting it is good to carry the
right foot behind the left and to continue with the left behind the right in order to rest on the right foot, for in
this way you will withdraw so far that your adversary cannot
hit, unless he has hit in counter-time. This guard of the
left foot will be more useful with the sword and dagger than
with the sword alone. But it is better to stand with the
right foot forward, and immediately after hitting draw it
back close to the left, for in this case if your adversary
follows you can advance it again, and also you can step backwards with the left, as you see an opportunity, hitting at
the same time as the adversary follows.

After these rules it is well to understand the pass, a
thing very profitable and advantageous, because you thereby
disturb and frighten your adversary, hit with more force and
show your superiority. The body, the sword and the feet are
more in union, and that union generates strength and vivacity
of movement. In the course of passing you can readily change
from one line to another, so that your adversary has difficulty in defending himself and has no chance to do much, since
the opportunity quickly passes, nor has he time to consider

much, and as his point is penetrated he cannot hit. In lunging it often happens that you find you have gone so far, either by carrying the foot too far, or because your adversary has himself advanced, that you cannot move out of distance and are hit in withdrawing. In such a case it is good to continue to the adversary's body, for the greatest danger is in getting within distance; but when you have passed his point and follow on to the body, you arrive before he can withdraw his sword. Still one often sees the adversary, though his point is passed and he hit, withdraw his sword and make a hit. This is due to the mistake of the one who passed, who has not gone right on to the body nor taken the _time_ well. For if he passes at the moment when his adversary advances his sword, or when his sword is occupied in the defence, or out of line, the adversary cannot withdraw his sword at the time of the pass.

We might add that, although you pass, you should still follow your adversary's sword gliding along his blade, wherever it is, so as to be continually defended. If he withdraws his sword, so much the better, as he uncovers more parts, and his _forte_ is drawn back and cannot resist. There are some men, who, although you have passed, do withdraw and make a hit; this is easier with short swords than with long. As to this we say that, whether the sword be long or short, if in passing you go close up to the adversary's body, you

will be safe. For in passing you can do various things, throw the adversary into disorder by jostling him, seize the hilt of his sword, and pass so far beyond his flank that he cannot bring his sword, however short, so far back without himself withdrawing, which he cannot do in time. In passing you can grapple with your adversary and throw him to the ground, which would be good when your sword has not hit; for it may be held for certain, that if in passing you hit, your sword would penetrate to the hilt, which would shake and disorder your adversary, and the wound would be inflicted in a part of such importance that he would be at least prevented from withdrawing his sword. Moreover the one who passes is in all cases the readier to seize a chance, than the other who is occupied with the defence and confused by the danger in which he is.

In addition to these advantages there are many other things to be done after a pass, which cannot be done when you remain steady. You can often in passing use the method of avoiding and turning the body; whereas if you do not pass or your adversary does not pass, you cannot do this so well. For if you wish to move your body out of the line of his point, either on the one side or the other, you can only do so by advancing; for two reasons, the first, that you may be able to hit at the same time, the second, in order that his point may pass before he can change its line again. When he

has penetrated so far, it is better to pass right on than to turn back and be hit by a second thrust before you are in safety. It is true that with the sword and the dagger it is more difficult to pass, and you must be more careful, since after passing the point of the sword there is the point of the dagger, and thus the danger lasts longer. Nevertheless there are rules for passing with safety, as will be shown when we deal with those strokes.

He who can pass well, is more sure with his sword, restricts his adversary more and is more certain of himself. Much judgment is needed in carrying the body and the feet correctly, so that the sword may perform its office. You must take care in passing with the left foot forward not to carry the left side forward, especially with the sword alone, as in that case you could not use the _forte_ of the sword, which would be too far back. Therefore even though the left foot goes first, the right side must accompany it. In this way you will be able to carry the body out of line, your sword will be stronger, and the point be as far extended as if the right foot had advanced, for the body can bend further forward.

To understand the lunge is one thing, to understand the pass another. With the knowledge of the two you can adopt whichever seems best according to your opponent and the circumstances. For sometimes you can lunge and cannot pass ow-

ing to the shortness of the <u>time.</u> This applies when you are in presence. For there is another kind of passing, which may be made in the least possible <u>time.</u> Its principles are different and will be treated of in another place.

48

ON HOLDING THE SWORD EXTENDED, STRAIGHT, AT AN ANGLE
AND WITHDRAWN.

There are various ways of holding the sword and the arm,
as will be seen in the following plates, which will illustrate
the variety of the guards. Since one method is better than
another we shall treat of the principal ones, reserving a ful-
ler discussion until we treat of the nature of the guards.
They will be illustrated separately on the plates. Some hold
the sword at an angle and the arm a little advanced towards
the knee with the hand in _tierce,_ or slightly outwards towards the guard in _seconde._ Others hold the arm withdrawn
and the sword in such a manner as to make a straight line
from the elbow to the point. Others extend the arm as far as
possible and hold the sword straight, making a straight line
from the shoulder to the point of the sword. This method is
very cautious, because it keeps the adversary at a distance,
but is very fatiguing, and the sword is weaker than with the
other guards because of the distance of the hand from the
body. In this position your sword is more easily engaged by
the adversary and great pains are needed to keep it free.
When you can do this, the position is a great impediment to
your adversary, because he cannot approach so as to hit, seeing the point so near, and cannot advance owing to the same

danger, unless he can engage the point and drive it out of
presence. Even though he places the _forte_ of his sword
against your _faible_ and tries to hit, it would hardly succeed,
since there is little uncovered and he cannot hit unless his
faible passes your _forte_, which you could easily prevent. If
he tries to hit below, he will easily be hit above, for your
sword being nearer and already extended must arrive first.
Therefore in order to hit more safely he must remove your
sword, and seizing the chance carry his body out of line on
the one side or the other and pass on to the body. For he
cannot hit until he has passed the point nor save himself or
his recovery; therefore it is better for him to follow on.
This method is the more likely to succeed, as it is difficult
for one who holds his sword thus extended and high to maintain
his point in line since with but a small movement his adversary could pass out of line. He could easily pass underneath
by lowering his body. It is however true that one who forms
this guard properly holds his sword extended and keeps his
feet close together, so that the lower parts are kept withdrawn, as they are more exposed and difficult to protect.
Also he can then advance further in hitting and similarly retreat, if his adversary approaches too near. For with this
guard the adversary must be kept at a distance, otherwise he
would find it easy to pass. For the same reasons the guard
is a good defense against cuts, since the _forte_ of the sword

is already pushed forward, to that the adversary's sword cannot fall without meeting it. If he tries to hit below, he
cannot reach before one, who holds his sword extended, has
arrived with the fourth part of his blade. If he keeps his
feet close together, he can reach all the further, although
the extended arm is in greater danger. Still it is easy to
defend by a slight motion towards the part threatened by the
sword, lowering the point more or less, as the cut is high or
low, and keeping the point in line. Certainly you should often practise this guard in order to learn how to hit without
hurling the arm forward. You must hit, but you must keep the
arm steady, and let the motion of the foot and the body suffice. This guard will teach you to hold your sword close to
the adversary, where you can more easily hit him, and similarly to keep it free. Some hesitate to advance the sword,
lest it should be engaged and subjected by the adversary.
You will learn also to hold the arm correctly, and after such
practice, when the opportunity comes, you will act more promptly and correctly. One who is unpractised often makes a mistake of too much or too little, and is not sure in his defence; moreover he does not extend so far as if he had practised.

Those who hold the sword at an angle in _tierce_ with the
hand before the knee, or in _seconde_ with the arm outside, have
a stronger hold of the sword, but the body is too much expos-

51

Your adversary can approach further, and with this _tierce_ you cannot disengage, as with your sword at such an angle it would take too long. In the _seconde_ although the sword is at angle you can easily disengage; but both of them are bad in defence against an opponent who can thrust in the straight line, because such thrusts come to the body without approaching the _forte_ of the sword held at an angle, so that in the effort to parry you would have to make a large movement and often would be too late. Even if you are in time the movement is so slow, that your adversary may easily change either into another straight line, or into an angle, as opportunity offers. For a thrust at an angle is most likely to pass, but thrusts in the straight line cannot pass one another; if of equal force they will nullify each other. If you hit it will be because your thrust was stronger by having engaged his _faible_ better. The weaker will always be driven out of the line, and the other will pass on and hit. But the thrust at an angle passes on and hits without a junction of the blades; such thrusts rather yield to one another, and therefore are very likely to pass and hit the part aimed at. Further one who fences with his sword at an angle can change only by a large movement. It is impossible that his point and hand should not make a large circle in the direction in which he has moved, all the larger if he changes from one angle to another, and incomparably larger if he disengages.

52

The movement, however, would be smaller, if the change is from an angle to a straight line, but would still be so large that, if within distance, he would be hit. To hold the sword at an angle is well enough for thrusting, but not for the defence. To proceed against such a guard with security it is necessary to be able to use the advantage not only of the sword, but of the body and the foot, and to realise well the strength of the angle, otherwise while hitting you will also be hit.

To hold the arm withdrawn and the sword straight, forming a straight line from the elbow to the point, is a better rule. In thismanner you can better acquire the superiority, hit and parry, and on occasion disengage more swiftly, since your body is more defended by the _forte_ and the point is more easily maintained in line. Still you should know how to use any method at need, for you cannot understand the nature of what you have not practised, nor to what it may lead. You must remember that one rule will not serve for all cases, but each has its appropriate end, and what is good in one case will not serve in another. Therefore, as we have already said, you must be rich in devices and understand the time when they may be used.

To adopt the safest position for the body and the best for the sword you must hold the arm not quite extended, but still rather extended than withdrawn, and the sword in a

straight line, or inclined slightly outwards according to the position of your adversary. In this manner your guard will be the best and your body safe with respect to the _forte_ of the sword, which can defend with little movement, as it is already advanced. Your sword will be stronger than with the arm fully extended, and in every case you are more master of it and can use it with more variety. It is less restricted and less fatiguing, nor is it so easy to pass under the guard as with the arm extended. You can change your position according to the occasion, and keeping the _forte_ always in its place you will defend with ease, if you use it in the proper manner. This guard is better than the other when remaining steady against your adversary, though our opinion is that you should remain steady in no position for long. Though you may be more secure than your adversary, yet all have defects. Therefore the judicious man, seeing his adversary steady in any position, will not only realise the fact, but know how to proceed against him and hit him. Also he will understand what such an adversary can do in attack and defence. But if he finds his adversary is not steady, he cannot so easily estimate the position, although from the first putting of hands to the sword and the manner in which the sword is carried, he will come to understand where to take advantage. Of this we shall treat in the second book, when we shall explain whether it is better to remain steady in presence and await a _time._

or to attack without a pause.

WHETHER IT IS BETTER TO CARRY THE BODY HIGH OR LOW.

To defend the body with ease you must know whether it is better to keep it upright or to bend it. You must consider that a body is very large compared to the blade of a sword, which though moderately long is still very narrow and insufficient to cover the body. The larger the body the greater the difficulty the sword has in defending it owing to the large movements required in defending the uncovered parts of the body.

In this matter some put forward certain reasons saying that the upright position is more natural, not so dangerous for the head, quicker for movement, less fatiguing and less restricted than the position when the body is bent. We say that some of these reasons are true, and some not. First because in the upright position you are in more danger, and cannot so readily attack, since you must make a great movement in defending the body and cannot make an extension without bending; in bending your movement is so great that you cannot recover in time. Being upright there is lack of union in your movements and less strength. Similarly your weapons are weaker.

When you understand how to control the figure and stand

without strain, to stand low is more useful. But if you cannot do this it is better to stand upright, for if you are in
a strained posture you cannot be quick to move. Whereas a
body bent at a suitable angle and well balanced on the feet is
much securer being less exposed, and can be defended with
little movement of the weapons. The forces are more united,
and this union generates vivacity and swiftness of movement
in one who is well used to this position. To take up this
position in the required manner needs practise and entails
fatigue. But afterwards it will be found more quickly and
easily; you will be readier and safer in every case, will
defend yourself without getting into disorder, hit more swiftly and reach further. The reason is that being already bent
the body goes forward without a great movement. Still in
this position you must take care to rest the body on one foot
only, so that the foot which has to move will be free and go
quickly, otherwise it would be late. You cannot lift the
foot without lifting the weight, and though it may appear
that you can do both at the same time, nevertheless such a
movement is slower. Further if you can work in union with
both weapons and body you will be in less danger as to the
head, since it will be nearer the _forte_, and you will be
more ready to pass on the one side or the other, and better
able to retire than when upright.

    If a man could make himself so small that his weapons

would cover him entirely, it would be well; but since this
is not possible, you must at least cover as much as possible,
and protect yourself, which will be equally good, taking care
to do so without constraint, so that you may work with agili-
ty in all cases. Thus the fatigue which you undergo in practising will be less than the benefit you derive from it. It
is a matter of defending both life and honour. He who can
conduct himself against his adversary with greater circumspect-
tion and security, will deserve greater praise and acquire
greater honour. It is clear that sometimes a great victory
depends on a small advantage.

ADVICE ON HOW TO PROCEED AGAINST TALL, SHORT, WEAK,
and STRONG OPPONENTS, and AGAINST THE CHOLERIC and
the PHLEGMATIC.
_____

In meeting your adversary you must subtlely consider not
only his character, but his strength and height, because you
must proceed differently according to the particular quali-
ties of your opponent. Therefore we will treat somewhat of
this matter and advise you as to the best method of holding
yourself. A tall man meeting a small man should recognise
his advantages; for being taller his reach is longer both
because of his height and the bend of his body. His body
goes so far forward that he can reach his adversary, while

his adversary cannot reach him. For this reason it seems he should incline to the attack, rather than to the defence. There is no necessity for him to gain the superiority over his adversary's sword, though this is good, but merely to keep his own free in order to hit when his adversary approaches, and before his own body is in danger. He must contrive to keep his adversary at a distance, for he must recognise that he is first within distance, and, when the shorter adversary tries to advance to get within his distance, that is the time for the taller to hit, or to throw his adversary into disorder by making a feint of having taken the time offered by his movement, and then hitting in the part he has uncovered by going to the defence; and immediately breaking ground, for then he will be out of the shorter man's reach. Or if he can do none of these things, it is well to play the adversary by breaking ground as often as he advances, never letting him carry out his plan, and to continue this until he finds a chance to hit or to put his adversary in subjection. He should proceed without passing, but merely carry his body out of the line when his adversary passes, or break ground so that he cannot pass, but is held and met by the point. In this way the shorter man will have to adopt another method, give up all idea of hitting, and defend, since he cannot hit without first bringing his body into danger. Therefore it is more useful and necessary for the short man to

gain the superiority and engage his adversary's sword in order to prevent a thrust, when he advances to gain distance. When he is within distance and finds a _time_ to hit, it is better for him then to pass, for he will have penetrated his adversary's point so far, that he will with difficulty get out again in time without being hit, unless the sword of the taller man is so far out of line, or so far withdrawn, that the shorter man clearly perceives he has time to recover and return to his guard. For he could not recover so far, in one movement without the sword of the tall man, owing to his length, reaching him.

On the other hand it is true, that, although the tall man has the advantage of the reach, a matter of great importance, his movements are slower and there are more uncovered parts, so that he cannot so easily get out of line. Offering a larger target he gives the small man a great advantage in hitting, when the small man knows how to conduct himself within distance. For the sword of the small man covers him more; he does not need to make such large movements in order to defend himself, and he has passed the greatest danger, that is penetrated the point, before his adversary has penetrated his. Since his exposed parts are smaller he is in less danger, and consequently his movements are safer than those of the tall man.

A strong man also has a great advantages over a weak man.

59

The basis of his procedure should be gaining the superiority over his adversary's sword, when he can easily drive it into disorder and hit while it is moving. For when a weaker sword tries to resist a stronger by changing from one position to another, it is driven out of line, so that the stronger man may easily hit. If he has lunged, he can recover, and repeat the effort. If the stronger man desires to pass, that too would be good, for when he came to grips he would likewise have a great advantage. On the other hand a weak man against a strong man must always avoid his sword, lest he engage. Nor should he parry, when the strong man hits, for it is often seen that the _faible_ of one man is stronger than the _forte_ of another; thus the weaker man would be deceived thinking to defend with his _forte._ Even where there were no great difference between the swords, when for example the one sword found not so much of the _faible,_ and the other not so much of the _forte_, since in certain cases a single hand has more power than two hands, it would still be better not to parry, if possible. For even if the defence were sound, the sword would receive such a shock that it would be very difficult to hit at the same time, unless the subtle method of reaching the body before the other touches your sword were followed. For in this ease the adversary's body strengthens his sword. The alternative for the weaker man is to avoid and free his sword, not to attempt to advance on his adversary. Also it is always

good for the weaker man to defend by withdrawing his body a
little so as to feel the shock of the thrust less in parrying;
though it is better to parry cuts by advancing, as the cut
has not the power of the lunge. He must keep his adversary
at the point of his sword, so that he cannot pass, remembering that to let him come to grips, would be the worse for him,
the weaker. If the heavier man passes and jostles him, he
would disconcert him so that he could do nothing, and before
he recovered the heavier man would have done many things.
Thus it is not good for a weak man to try to acquire distance,
but to contrive to keep out of distance, not to let his sword
be engaged, but on different occasions to incite his adversary
to hit by offering a time: or to make a feint of offering his
sword, so that the adversary may believe he can engage, and
whilst he is moving, break ground a little, and thrust at the
part uncovered by the heavier man, so that he may be hit as
he comes on, and, thinking he has passed, find the weaker man
is away by having broken ground, so that he cannot reach him.
The stronger man will thus be disordered and may be hit before he recovers, if the weaker manhas not already hit and
brought himself immediately back into safety, with the intention of letting his adversary's sword pass in vain, saving
himself with his body and feet. Therefore for these reasons
it seems clear that it is ill for a weaker man to attack a
tronger man; it is better for him to stand on the watch

with the idea of defending himself by the advantage of distance.

Again in dealing with a choleric or violent opponent you should endeavour to provoke him to attack, so as to hit him when he comes. It would be ill to seek to approach him and come to grips without the advantage of the point; but rather you should encourage his fury, giving him opportunities in order to cause his downfall more easily; while he is approaching you can advance, or retreat according to circumstances, in order to defend your self and hit at the same time, and before he passes. On the other hand, if you have to deal with a phlegmatic opponent who waits, then you can attack; but take care you are not deceived, for often in the desire to hit and the belief that you have merely to frighten your adversary to the defensive, you are yourself hit; whereas if you wait and are restrained, you may easily defend and hit. Therefore you must always consider the danger, with whomsoever you are matched; you must never rashly despise your adversary, but always be on the watch for whatever may happen and ready for all accidents.

Our discourse so far has been to show the principles on which the science and practise of the sword are founded. We have omitted many things which we might have said, and have had regard only for what seemed to us more useful and necessary, and more in accordance with the use of the present time.

62

Nowwe shall treat of the nature of the guards and movements, illustrated by the plates. In each guard the illustration will be double to show the position of the right and the left side of the body.

GENERAL DISCOURSE ON THE GUARDS.

We have now reached the point when we must treat of the formation of the principal guards, and movements and the results obtained in arms. We must first warn the reader not to wonder if he sees two figures illustrating one result. This is done to represent the right and the left side of the body. On the other hand we have thought it unimportant and idle to treat of many other guards of which some authors have written; for instance a guard with the dagger extended and the sword thrown behind, now on one foot and now on the other, now high, now low, which seems to us to defend the rear rather than the front. Others with the sword alone have kept it so far back and low, that the point was near the point of their feet, and also they held the sword across the legs and with the point almost on the ground, and all this that the sword might not be engaged. Sometimes on guard they take the blade in the left hand to give it strength, in order to beat the adversary's sword and hit. All these things we have omitted as inappropriate and, more often harmful than useful, and in any case tedious to the reader. Perhaps it had been better

to have passed them in silence, but some might have thought
we had not seen or considered such things;- therefore we have
made some mention of them, as of the practice of throwing the
sword at the adversary, when fighting with the sword alone;
some think this an essential movement, but we deem it of little value; it may succeed against those who leave the sword
free or hold their own too stiff, but against those who engage their adversary's sword and can disengage, it effects nothing, rather he who adopts this method will always be beaten.
Therefore we shall not treat of it further in the present work,
but shall try to give such discourses, as when well considered
can bring you such counsel and judgment, that, when you see
your adversary approaching sword in hand in whatever manner,
you will recognise the principles he is following as well as
he himself. These results are illustrated by plates, from
which you may expect great benefit. To these are added the
discourses not only as an explanation of the results, but also in order that you may discover the intention of one who uses
them and so anticipate your adversary's thoughts and prepare
yourself before the result follows.

Similarly the reader must not wonder seeing these exten-
sions of the sword, feet and body; they are merely to show
how you must proceed, when on guard, when passing, parrying,
and hitting. Some swords will be long and some short; tall
and short bodies will be seen, according as they are held more

or less low, or as they are held upright or bent. These
things will explain the guards formed both in the defence and
the attack, the position of the body and the movements to be
made, the one differing from the other according to the occasion. After these simple illustrations will follow others, in
which will be revealed the parries and the hits which may
arise from them, and their causes will be considered; you
will understand that all movements of attack and defence must
be made in _time_, when you have the sword alone in your hand.
Other plates will then follow with short discourses;   where
necessary we shall discuss them at sufficient length, but
where it is not necessary we shall leave them to the consideration of the reader. In these cases we shall explain merely the position from which the hit followed, how it is defended, and what was the guard formed before this result. In
short we shall try to give such instruction, that you may easily learn what to do in any position, when you encounter an adversary, both what your adversary can do in attacking and how
to defend, and similarly the changes of line which may be made,
and in what distances, wide or close, of these distances we
shall treat in different places, now of the one, now of the
other, in order that you may understand within which distance
a stroke is made.

65

DISCOURSE OF THE PLATE SHOWING THE NATURE OF THE CUTS
AND WHERE THEY HIT.

This plate shows the nature of all the cuts, which a
hand can make. The names are placed against them so that you
may see where each of them naturally hits, although they may
hit higher or lower according to whether they are made with
the hand or the arm. At least their path is seen, and from a
knowledge of that follows a knowledge of the second point,
what sort of defence can he made in order to parry them and
hit at the same time. Therefore the names on the plate are
placed not in the part from which the cuts are delivered, but
where they hit; for the cut of _mandiritto_ is delivered from
the right and hits the adversary's left shoulder, and the cut
of _riverso_ is delivered from the left and hits somewhere on
the right side, as may be seen. Whoever examines and ponders
on these cuts, will easily discover the principles of proceeding against each one of them, bearing in mind that even if all
the cuts are made by the same arm they may not have the same
strength, and therefore against the stronger it is necessary
to find a stronger defence in order to resist and hit. Although
it might appear that we should here treat of the differences
in the cuts, still we think we have treated of them sufficiently in speaking of the defence and the attack, and of thrusting

and cutting. It is our intention to base our instruction,not on these, but on more subtle and profitable principles.

PLATE I.

DISCOURSE ON THE GUARD IN PRIME FORMED IN DRAWING
THE SWORD FROM THE SCABBARD.

———

This plate (the second in order) shows the position into which the hand goes in drawing the sword from the scabbard, wherefore its name of the guard in _prime_. It cannot be held to be very secure, since the sword is too much withdrawn and the body entirely exposed owing to the height of the sword, which brings the _forte_ very far from the body, so that it cannot defend the exposed part below in time. In this case you would have to defend with the hand, unless you broke ground, otherwise you would be hit before parrying. If you wished to hit after the parry you could lower the point a little, abandoning his sword, and make a cut, or a rush, but as this would be a hit in two _times_ it would not be very successful. As to the head it is sufficiently defended by the guard, and more on the outside than the inside. But we shall form another guard which is safer, with which you can await your adversary or advance. With the present one an advance would be very dangerous; therefore this position of the body and the sword should be used on breaking ground rather than at any other time.

67

PLATE 2.

EXPLANATION OF A WELL FORMED GUARD IN PRIME.

If you wish to form a sound guard in _prime_ the position of the body and sword must be as shown in the plate, the feet close together, body bent, arm extended, with the sword in front with the point as straight as possible; for the point will naturally incline towards the ground. Thus your adversary cannot thrust over the sword; this part being the weaker must be better defended. Moreover you must keep the feet together and the body bent, so that the lower parts may be so far withdrawn that the adversary cannot reach them without penetrating with his head half the sword's length. Your sword will have to defend only the head and part of the chest, which it can easily do, as the _forte_ is already so far advanced that the adversary's sword can never extend so far as not to be always nearer the forte than the body. This guard is excellent against cuts, for with it you can defend and attack without turning the hand. It would be as good as any other guard in fencing if it were not so laborious for the arm, that you cannot long endure this position. With this guard you can advance to engage and harass your adversary's sword without changing guard, always approaching so as to hit on the outside

over the sword, or below, in case your adversary disengages,
by lowering your body still further, moving the feet apart,
and keeping the arm in the same guard; as soon as you have
hit, bring the feet together again and try to engage his
sword above, even though his sword is on the inside, and push
it out. This you can easily do, for your adversary cannot resist, as in this line your guard is strongest.

PLATE 3.

THE GUARD IN SECONDE, ARISING FROM THE GUARD IN PRIME
FORMED ON DRAWING THE SWORD FROM THE SCABBARD.

From the position of the hand in drawing the sword from
the scabbard arises this guard, with the arm somewhat lowered and turned downwards. This has caused a slight change in
the front of the body. It is called the guard in seconde
because it is the first movement which the hand can make in
changing from the guard in prime. It is easier than the guard
in prime, as the arm is not so strained; owing to the change
of the position of the hand the weak part has changed. In the
first it was above, now it is on the outside. It is true that
as the feat are rather far apart the leg is in some danger towards the knee; still if you can keep your sword free, your

adversary will only with difficulty hit you so low before he
is himself hit above. Although in this guard the arm is somewhat withdrawn, the _forte_ is so far forward that it can parry
excellently both on the outside and the inside; but the hand
must be turned in _quarte_, or you must parry with the hand. If
the feet are kept closer together this guard will be safer on
both sides. But we shall form another guard like the first
but much better.

PLATE 4.

A WELL FORMED GUARD IN SECONDE.

This is the position in which you should form the guard
in _seconde_ for greater safety. Although it is fatiguing, it
is less so than the guard in _prime,_ for the arm is somewhat
lower. Its weakest part is on the outside, therefore you must
hold the point so straight that your adversary cannot come in
there. Although it is the most covered part, only a little of
the head showing above the right arm, your adversary might
come in there, put you to the necessity of defending the spot,
and then proceed to hit below. If then he should attack on
the outside you should disengage, without advancing, if you
have not been able to hit whilst he was moving on the outside.
The lower parts are still more secure than with the guard in

prime. The defences are somewhat different, for you must
defend the mandiritto tondo by turning into quarte, as also the
sottomano. All the others are parried with the guard/seconde ^{of}
except some thrusts on the inside, which are likewise parried
with a turn to quarte. You will do well with this guard, as
your sword is advanced and straight. If you understand its
principles you will find it excellent and advantageous; it
leaves little uncovered for the adversary to hit, and the
body is so far withdrawn that he cannot reach it without first
subjecting your sword, which is difficult as the disengage is
with this guard made with little movement and quickly. But
as we have said above it is somewhat laborious to maintain
for long.

PLATE 5.

PRINCIPLES OF ANOTHER GUARD IN SECONDE FORMED WITH
THE ARM FORESHORTENED AS IN THE PLATE.

———————

This strained position is a guard in seconde. In spite
of its appearance it has great swiftness and power, because
of the union of forces. You begin to form it when upright,
as the adversary approaches so you lower the body and withdraw the sword in such a manner, that, when within distance,
you are so low and the sword so far drawn back, that it can

be taken no further and the point still kept in, nor the
body further lowered. Your sword must form a straight line
from the hand to the point, so that the adversary cannot proceed on the outside. Also you must keep the hand to the front
to defend yourself from any rush which might he made before
you had completed the guard. When completely formed and your
adversary approaches so far that his point just penetrates
your point, if your sword is free, you must quickly change
into quarte on the inside. For this purpose you keep your
right foot so far across, that, when you lunge the body goes
out of line before the feet move; in this way your lunge will
be longer and pass right to the adversary's body. But if the
adversary's point is directed towards yours, you must rush
with your body under the point, thrusting in seconde past the
faible of his sword, and pass right on to his body. If you
see that your adversary's sword so far impedes yours that you
cannot hit in that line, you must disengage from seconde,   .
leaning your left hand on the hilt, so that the adversary's
sword cannot push yours aside, and hit in the upper lines on
the outside.

Cuts are easily defended in secondo or in quarte according
to the part they are directed against. These rules will succeed very well against those who do not understand the foundation of this guard. He who uses this guard, when making a hit

will generally parry with the hand, and in all his hits will pass with determination. The guard is fatiguing and confined, still one who is well practised in it can effect much.

PLATE 6.

In this plate the sword is shown foreshortened, and the left side as far forward as the right. You have formed a guard in _tierce_ and changed into _seconde_. The sword is turned so far to the left as to be quite fore-shortened, and therefore only the cross or hilt is seen. This movement has been made in order to let the adversary approach. The body is bent forward so that it may not be hit save on the head and chest, and if the adversary attempts to hit you can parry with the left hand, which is held before the face, hitting with the same movement of the body and extending your sword into _seconde._ If you have completed the position when your adverysary advances, you can change to _quarte_ and hit below or above his sword, according as he comes low or high, and can carry the body out of line without parrying, though you may parry and hit with this _seconde_. If your adversary does not respond to this _appel_, you must not remain in this position, but change your line, remaining steady on your feet, so that he may not take the _time_ on that change; for if your feet were being brought together you could not parry, but if you were

withdrawing it would be well to parry, since your adversary could be sure of making a hit. If when he took the <u>time</u>, you were steady, you could advance or retreat according to the distance and intention of the adversary, because you would have adapted yourself for attack or defence at the same time.

<u>PLATE 7.</u>

ON WHAT OCCASION YOU MAY MAKE THE EXTENSION IN <u>SECONDE</u> SEEN IN THIS PLATE.

The extension seen in this plate is made in <u>seconde</u> with the right foot, and can be made on the inside or the outside of the adversary's sword in the time when he is passing. The lunge is made with the idea of letting his sword pass in the air without parrying, if, as might be, he is in a guard of <u>tierce</u> or <u>quarte</u>; but if he is in <u>seconde</u> you will not succeed with this lunge. If the adversary does not pass, it is not a good movement, for the body is so low and the feet so far apart that you cannot recover quickly enough to protect yourself. You should certainly make this stroke if your adversary passes in order to save yourself from the impact of his sword without parrying, and hit him at the moment of his passing. If you realise the opportunity it is quite safe, be-

cause the body is so low, that the knee and the head are covered under the line of the arm in such a way, that even if the adversary attempts to hit at the centre of your body, he will pass far above. Thus the position deceives the adversary; but you must have a care not to form it at too great a distance, for he could then lower his point again before it had passed, and your head would be in greater danger than before. If the movement is made within the proper distance this danger ceases, for at the moment when your adversary's sword approaches, your body goes to meet him and causes his sword to pass with even greater celerity.

PLATE 8.

How to make this extension in seconde by advancing the left foot in order to make a long extension.

This extension is made with the guard in _seconde_, with the left foot brought forward in the time of the adversary's movement. Here you may clearly perceive how low a body may go and pass right to the adversary's body with swiftness, when you know how to control the body. This figure is drawn from life like all the others. With this manner of advancing the foot the lunge is very long. The body is protected by its low position. The left foot is carried forward, as is seen, but the right shoulder and side also go forward in such a

manner that the lunge is as long as possible. The occasion
for its use may arise, not only when the adversary tries to
lunge or pass, but at the slightest _time_ offered, when you
are within such distance, that at the first movement you can
pass the adversary's point. You may take the _time_ to make
such a lunge whether on the outside or the inside, for the
body is so low that your adversary's sword remains far out
of line, and all the more when his point has been no lower
than your chest. Further this extension is made with great
velocity, much more so than the first extension.

PLATE 9.

This plate illustrates the guard in _tierce_, which arises
from the guard in _prime_, as does the guard in _quarte_, as will
be seen. The guard in _tierce_ is less fatiguing than the other
two, because the arm is in a natural position. But the hand
is too low and the point inclined upwards at an angle, so that
there are many exposed parts. If you draw a straight line
from the point to the body you will see how great is the angle,
and all the space between that straight line and the hand is
exposed, where you may be hit on the outside and on the inside.
On some sides it is not strong; further you may be hit before
your hilt is reached. If you wish to defend the upper parts
you make so large a movement with the hand, that you cannot

reach the defence in time, and your adversary can easily deceive you. Again, as your sword is at such an angle, in parrying you often go out of line, offering a great advantage to your adversary. Again, by bringing the sword into the straight line you weaken it, for the sword is always weakened by being extended, with great danger that your adversary's sword, if already extended, will remain the stronger, since the sword is always stronger after coming to rest than in its passage. Further, one who stands in this manner with his sword at an angle, can make little use of the disengagement, since his point has to make too great a circle and too large a movement. Nevertheless this guard may be used, because its nature is not always known. Though the sword is at a great angle and the body much uncovered, still you can sufficiently deceive your adversary by avoiding and freeing your sword by a half disengagement. Therefore he who can use these devices in time will defend and easily protect himself. The desire of the adversary to hit is increased by the sight of so much exposure, with the result that you may readily save yourself and hit your adversary. But we shall form a safer guard in tierce.

76

PLATE 10.

This plate shows the manner of forming a sword guard in _tierce_. The position of the arm and sword requires exactness for the hand should not be turned at all. Where the natural guard in tierce is weaker, this guard, also in _tierce_, is stronger. By the change of position the posture of the body has been changed to great advantage and improvement, since the angle of the side is well drawn back. The outside is defended and there is little uncovered on the inside. This is the true method of trying to engage the adversary's sword whether on the inside or the outside, for there is little or no movement to be made with the arm, but with the point of the sword only, which will so subject the adversary's sword as to keep it always underneath; your hand will then be between _seconde_ and _quarte_ in such a way, that you can engage with little movement in the one or the other line according to opportunity. Therefore we consider it to be one of the best guards.

77

PLATE 11.

We have wished to include this guard in tierce, since
it has some advantageous principles, as we shall explain. It
may be derived from the extended tierce if in that tierce the
sword is in danger of being engaged, or for any other reason.
You will free yourself by disengaging from that tierce to
this, for the inclination of the body is now backwards, as
may be seen; without moving the feet, but by merely bending
the body and the knees the body is carried so far back, that
the adversary cannot hit. At the same time the sword is freed.
If your adversary attempts to engage it or to hit by carrying
himself forward, with this guard you may very well hit him
simply by again advancing your body forward at the same moment
as he advances. Further it is a sufficiently good guard to
practice, because it forms an oblique line towards the ground
in such a manner, that your adversary cannot easily seize it.
If he tries to engage your sword so low down without using
the same advantageous position of the body, he will certainly
be hit, since the distances are very deceptive. When within
distance you appear to be far out of distance, and when the
part which is bent backwards is bent forwards,without moving
the feet, you extend more than half a sword's length by the
mere inclination of the body, so that you reach further than
your adversary estimates, if he has not understood the nature

of the position. Thus as this guard may serve for this long
reach, it may also serve for a withdrawal, since the distance
is enlarged more than half a sword's length, with the result
that your adversary cannot reach in _time,_ and you have the
advantage in defence and attack. But you cannot engage his
sword before coming within close distance, unless you are
careful to bring your feet together and bend the body as far
forward as possible; then indeed you may reach his point
though within wide distance. With this guard you should remember that the exposed part above is so far back that it cannot be hit if you know how to keep your sword free. Thus this
position is very much to the purpose and good against various
guards at an angle, and even extended guards. A body in such
a position could easily and swiftly get out of line of the
adversary's point, and with equal celerity pass on the one
side or the other, except against the guard in _prime_, which
could engage your sword down to the ground.

79

PLATE 12.

This plate illustrates an extension in _tierce_, which shows how to advance the foot, bend the knee and incline the body sideways, so that there is little exposed. It also teaches how to stretch the wrist in order to lengthen the line and reach further in a lunge, and to recover after making the thrust. If you wish to recover quickly, you must not straighten the body, but bend the knee of the leg to the rear, bringing the weight of the body on to it low down, so that the forward leg is straightened and relieved of the weight in such a way, that you can easily and conveniently raise it. You perceive that all these things must be done in one _time_, and that if you do not know how to shift the weight of the body you cannot readily raise the foot, especially when the feet are far apart and the body much bent, except with great inconvenience, difficulty and length of time.

80

PLATE 13.

This is a guard in quarte, the last of the four guards.
It is formed simply with the arm at an angle, for which reason you are strengthened, and sufficiently covered on the inside. But you cannot disengage with much promptness, and on
the outside are much exposed because of the angle formed by
the arm and the hand. Although your position is strong, still
if the adversary attempts to hit, it would not be so good to
parry or to attempt to hit under the sword on the right side
of the guard, letting his point pass without parrying. In
this manner you would be most successful; for your arm being
at such am angle, if your adversary wishes to reach the body,
his point must pass inside that arm. If the arm is extended
towards the inside, it will cover all the part, which was
seen exposed before, and will hit without touching the adversary's sword. With this guard, if you have an opportunity to
change from quarte to seconde, the result will be effective,
and still more so if the arm is fully extended from its present
angle. The hand will be so far on the inside, that the adversary's sword cannot cover much in that line. When you change
to seconde, an angle opposite to the first will be formed, so
that your adversary cannot be in as strong a position. It
will be all the better if in changing you continue on to the
body. For if you measure the path taken by your point in

hitting and the path taken by the adversary's point in defending, you will find that the path of the defending point is the
greater. Although your hand has moved from one angle to another, a large movement, still the point will have gone to the
body without shifting. Thus that large movement will have
done damage to the adversary and not to yourself, if you have
changed the front of your body by moving it out of the line
of his point.

82

PLATE 14.

The plate represents an extended guard in quarte. It
is much better than the preceeding one, and beyond comparison
more cautious than all the others, because it keeps the adversary at a distance. He cannot be sure of engaging your sword,
because with this guard you can disengage easily and subtly
and with greater promptness than with the other on one side
or the other. Its greatest strength is on the outside, where
you cannot be hit, and the inside is assured by the effect of
the hand being turned in that direction in such a way, that
no path is left for the adversary to take, except by pushing
your sword out of the line, a dangerous plan, as with this
guard the disengage and double disengage are swift. His only
resource is to try to disorder your sword by a feint or a
movement, in order to hit below, carrying his body suddenly
forward; for he could not reach with a lunge without putting
himself in greater danger of an attack. Thus this is the
most secure of the four extended guards, as we have said.
There is no other which can be maintained with greater ease,
with the sword more ready and free.

83

PLATE 15.

This is also a guard in <u>quarte</u> and very different from
the two last. As you may see in the plate, the chest is exposed to the adversary and the feet in an oblique line. The
intention is to move to either side according to the opportunity. The adversary cannot proceed to hit either your
chest or head, for your feet are on either side of his sword,
so that by lifting one of them your body will be out of line,
so that you can hit in <u>quarte,</u> in <u>tierce</u> or in <u>seconde,</u> as
the <u>time</u> and the occasion demand. With this guard you are
uncovered on the outside. You invite your adversary to attempt a hit there, knowing that it is the strongest part, and
that the angle is naturally of such a kind, that if he tries
to hit in that place, by carrying the left foot in a straight
line, extending the arm and leaving the hand in the same position, you will hit your adversary below on the right, or
above by making the angle still larger and carrying the hand
as high as the shoulder. In this way your lunge will be so
strong that, however much your adversary tries to parry, he
will still be hit. If he approaches too close without resolution, you should turn your hand from <u>quarte</u> to <u>seconde</u>,
covering the head and carrying the left foot forward, and
pass on with body and sword you will make a hit in the chest
in seconde. With this guard you must take care to be so

far advanced that as you change to _seconde_ the head may penetrate the adversary's point with the bending of the body, and
you may then proceed to hit and carry your left hand to his
hilt, if the adversary disengaged in order to hit in the lower lines, he would effect nothing, because your sword, which
would have already begun to change into that line, would prevent him, and would hit on the outside, for you would have
brought both sides of your body equally forward; this excellent result would be due to the length of your reach and the
strength of your sword. In this way the only difference would
be that your body would pass on the outside instead of the inside. With this guard you may easily use the left hand.

85

PLATE 16.

Although this guard resembles the last, nevertheless it has considerable differences, for in this guard the sword is held in such a manner, that the hand does not form an angle on the outside, but the wrist is inclined inwards, and bends the sword so that it appears foreshortened. Whereas the other quartes are stronger on the outside, this quarte has more power on the inside owing to this foreshortening of the sword, and also to the position of the body with the left side so far forward, as is seen. We have included this guard in order to show how you may advance on the inside to subject the adversary's sword, which is in seconde at an angle. You must know that the greater the angle your sword forms, the greater is your force in that line. If the adversary tried to hit your exposed part, he might himself be hit through the angle of the guard in seconde, if he did not hit with the advantage of the line for the straight line always reaches further, or did not avoid with his body, or wait until his opponent came within distance in order to be able to hit and withdraw. But for passing at the fitting opportunity without being hit there is no sounder or better position than the one seen here. For, whatever the angle of the adversary's sword, this foreshortened sword will push his way and will be stronger than the seconde, so that he will be forced to change his guard, or retreat; otherwise you will at the slightest move-

proceed to hit and pass to the body.

PLATE 17.

Here we have placed an extention in <u>quarte</u> to be used in lunging in order that you may understand the working of the foot, body and hand together. With this extension you hit with the head covered and without turning it, as some do. It would be better to turn and raise the hand a little, and if you had first extended in <u>quarte</u>, the hand would have moved little; nevertheless however great or small its movement, it would certainly be better than turning the head and losing light of your adversary's movements. Holding the head back with the idea of its being safer, is an obvious error, since the further your head is from your <u>forte</u>, the greater the danger it runs. Moreover, if you hold the head back, you cannot lunge far enough to reach the adversary's body. Also you must keep your eyes on your adversary's sword hand, not only when steady, but in passing and turning the body, on whichever foot. As to recovering with the body bent so far forward and the feet so far apart you must observe the rule described when we spoke of the lunge in <u>tierce.</u>

PLATE 18.

Here follows another extension in <u>quarte,</u> in which the right foot has been turned. The lunge is made by advancing the right foot only, turning it in the air in such a manner that the turn is complete as it reaches the ground, as you may see. The plate shows that the turning of the foot most begin with the lifting of the body, so that as you advance all the part which was visible when on guard is taken out of presence and the adversary's sword passes in empty air. It shows that you must hit the adversary at the same moment, recover the right foot at once and return on guard in case he has not passed or had not meant to pass. After completing this stroke you could follow with the left foot turning it backwards and continuing to the adversary's body, if he had not passed, as we said; for if he had passed the first stroke would have sufficed.

87

PLATE 19.

This is another extension in _quarte,_ made with a turn of the left foot. It may be used in the _time_ when your adversary tries to hit on the inside in _tierce_ or _seconde_, or to pass below. You can similarly make use of it if he offers a _time_ when on guard. But it must be accompanied by a movement of the feet, so that he cannot break ground, while you are trying to hit. Otherwise you would be hit, as you would also if you gave your adversary time to change line. For in turning in this manner, if your first plan fails, you can form no other. Therefore you should not turn unless you are certain that you are so far advanced, that with the first movement of the feet the body can pass the adversary's point, for otherwise you would be easily hit in the back. After turning the left foot it is well to follow right on to his body in order to get entirely out of presence and in order that he may not withdraw his sword and return to hit you. If your adversary passes it will be unnecessary to do anything but the turn in order to get out of line and escape the impact of his sword. In this case you would be more successful, as you could not be deceived.

88

PLATE 20.

This extension in quarte, advancing with the left foot, shows how the foot should be carried and the body inclined in order to make the lunge as long as if made with the right foot; it shows too how to carry the body out of line, so that the sword may defend better. This is certainly a better method of attacking, because you can change from one line to another as you advance, and form a new plan. But it requires a good knowledge of weak and strong points, because it is not a question of avoiding with the body, but of dropping under the adversary's sword. Moreover, in any case you defend with the forte of the sword according to opportunity. This method of hitting is more vivacious and the sword is stronger than in any other method of hitting in quarte.

89

PLATE 21.

This plate illustrates the first hit in _quarte_. It is a lunge against a guard in _tierce._ It may arise as follows. The adversary, who is in _tierce_ has made a feint of hitting on the inside, whilst you also were in _tierce._ He has come forward in order to make you parry. But you have taken the _time,_ carried your hilt up to his point and driven on your point to make a hit. By advancing the right foot, bending the body and turning your hand into _quarte,_ you have encountered and hit your adversary, as may be seen, at the moment of his coming forward. He has not been able to parry, while his foot was in the air and he was advancing. Similarly it may happen that you both are in _tierce_ on the outside, and the adversary has tried to disengage to the inside line, advancing his sword and body to force you to parry, with the intention of hitting you in that _time_ by changing from _tierce_ to _seconde,_ and lowering his body; or he intended to return to the outside in _tierce_ in order to hit over the sword. Both these methods would have been effective, if you had done what he desired. But you, being steady on your guard, with your sword free, within wide distance, were awaiting the _time_ in order to hit or seize some advantage. As soon as you saw the movement of his sword and body bringing him forward, you realised that although he had not moved his feet, he still could not break

ground, since it is impossible to advance and retire at one
moment. Hence you may realise the great danger of moving
without a _time,_ in order to advance, when you are already
within wide distance, especially when the adversary's sword
is free. If you are forced to move in order to free yourself
from some danger it is better to retire than to advance, especially if your adversary is steady on his guard, and thus
free yourself in such a manner, that if your adversary attempts to hit in that _time,_ you can defend yourself and attack at the same instant. Even if you have moved your sword
and body, provided that your feet are steady, you can always
save yourself when within wide distance. But within close
distance the smallest movement involves great danger, as we
showed in treating of distances and _times._ Further you must
consider that, while your adversary's sword is free and steady,
to make a feint, in our judgment, is merely to hurry fruitlessly. If your adversary makes a feint, he can never hit,
even though you parry, as long as you are steady on your feet,
If after a feint he tries to hit, you can break ground, so
that he will not reach, and will be thrown into disorder with
danger of being hit before he recovers. Therefore to make a
feint you must await some movement of your adversary, or not
leave his sword free, but engage it first, so that he cannot
hit in that line. Afterwards you can feint, without abandoning your advantage. In making this feint you must go forward,

so that if he does not parry, the feint will hit, and if he
parries, you can by a change of line reach his body, before
he can save himself by breaking ground. For the one who has
moved with the feint will arrive more quickly than the one
who has awaited the second _time_ in order to break ground. If
when you make the feint, your adversary breaks ground, you
must not advance, for you are too far away. You must stop
and return to the engagement. If the one who has been hit in
the present case, had observed that principle sooner, he would
have been the hitter rather than the hit, or would at least
have saved himself. We have made a long discourse on this
present hit with respect to its advantages and dangers, and
not only that, but the manner of its arising, and how else it
might arise, and how he who was hit might have found safety.
For before the attack many remedies were possible, though they
were of no avail after the thing was done. But in the following hits we shall state only the cause and the effect, leaving the rest, lest the reader should be wearied.

92

PLATE 22.

The second hit is in _tierce_ against a guard in _tierce._ It may arise when you finding yourself on the inside, have made a feint in the straight line and your adversary in parrying has dropped his sword through not meeting yours, for you have disengaged in the _time_ of his parrying. You have gone on to hit on the outside through the angle naturally formed by the hand in _tierce_, and the adversary has been unable to push your sword away as his _forte_ was already so far advanced that his sword remained locked. Or it might arise in this way: both being in _tierce_ on the inside, you have advanced to engage his sword. He has tried to disengage, advancing his right foot. In that _time_ you have pushed on and made a hit before the disengage and the movement of the foot were completed, in such a manner that his point has been pushed outwards, before he could return it into line. It is obvious that the _time_ offered by the disengage from one line to the other, being a larger movement, is longer than the _time_ offered by one who remains in the centre line and goes straight on. Therefore, you may say that you have arrived before he has finished the disengage, and in this manner have pushed him out of the line as the plate shows.

93

PLATE 23.

This is a hit in _quarte_ against a tierce. It has succeeded because both were in tierce within wide distance, and
you have moved your sword and tried to engage the adversary's
on the inside. He, seeing your plan and that you were uncovered below the sword hand, has lowered his point in order
to hit in _tierce_ in that line. But you, who have moved the
point only, seeing him coming in below, have abandoned the
attempt to engage and directed your point at his body, turning your hand into _quarte._ By carrying the hilt to his
faible you have parried and hit at the same moment. This has
been due to your adversary's ignorance, for he has failed to
realise that your movement was so small, that he could not
arrive before the _time_ was finished. Therefore he should
not have gone on. But it was good to lower the point in this
manner, if only he had not advanced or moved his feet. For
then against this hit he could readily have defended himself
and attacked in various ways.

94

PLATE 24.

This is a hit in _tierce_ against another _tierce_. Both
were on the outside, and you have moved to engage your adversary's sword. He, seeing the time, without considering the
width of the distance, and that you have moved the point only,
has advanced his foot in order to disengage on the inside and
hit in _quarte,_ or in order to engage your sword, and has advanced within close distance. You had moved with the intention merely of making him move. Seeing that he was beginning
to drop his point to disengage it and that he was advancing
his foot, you also have let your point drop; lowered the body
and have met his _faible_ with the hilt and prevented his disengage. At the same moment you have pushed on with the right
foot and hit under his sword on the outside. Therefore it
should be held as a true principle, that when your adversary's
sword is free and he is steady on his feet and has tried to
engage , you should not allow this danger to arise, but should
form a plan to acquire some advantage without advancing, but
rather retiring. For the movement of the foot is longer than
the movement of the sword. But in the case when you have engaged your adversary's sword and he has tried to free it, even
without moving his feet, then you can advance a foot and engage on the other side, in order to hit when he moves again.
Therefore the principle of advancing when your adversary moves
rests on the advantage of having first engaged his sword. If
it is free, it puts you in danger, as the plate shows.

95

PLATE 25.

This represents a hit with the guard in _prime_ against a _tierce._ Both the combatants were in _tierce_ on the outside, and have attacked each other's swords and have begun to push one against the other. You have turned your hand from _tierce_ to _prime._ This guard naturally tends to hit towards the ground and is strong in the lower lines. In this manner, as you have raised your hand, your adversary's sword is left below in your strongest line. Further in raising the hand you have approached his _faible_ and at the same moment have pushed on and made the hit shown. You could have made this same attack and with less effort, if his guard had been in _seconde_, since, although that _seconde_ covers the part hit better, still it is weaker, and therefore the _prime_ would have overcome it more readily than the _tierce._ This hit arises out of a tussle with the swords. In this connection you should bear in mind that it is bad to resist one who forces your sword; it is better to yield and release your sword, since the sword which tries force is bound to fall a great deal, when the other sword yields, which gives you a _time_ to hit or seize some advantage over the falling sword. This is a safer and less fatiguing method, as we have said.

96

PLATE 26.

This is a hit in _quarte_ against a sword in the air. Both were on the outside and you have tried to engage your adversary's sword, which may have been in _tierce_ or in _seconde_. He has raised his sword to make a cut of _mandiritto_ at the head. You were already moving in _quarte_ in order to engage his sword and have merely lunged forward, advancing the right foot in such a way as to arrive before his sword fell. Even if his sword had fallen, it would have effected nothing, because in lunging you have raised your hand far enough to protect the head from the cut. As the cut is made from the elbow, it has left much uncovered, is slow in hitting and has led to this hit. If the cut had been made from the wrist, in the _time_ of your hitting the sword would have fallen on yours, so that both the attack and the defence would have been illustrated together. But as that did not happen, the attack only is shown here.

97

PLATE 27.

This is a hit in _seconde_ against a sword in _tierce_, whose point is out of line; it has arisen in this manner: both were in _tierce_ on the inside and you have tried to engage the adversary's sword. At that moment he has turned his sword to make a cut of _riverso_ through the uncovered part seen outside the sword towards the head and shoulder. Seeing his sword making a circle you have changed from _tierce_ to _seconde_, covered the threatened part, and in the same _time_ driven on your point, arriving before his circle was finished. Even had his sword fallen first, he would have met the parry. All this arises from the slowness of the cut, which you cannot make without getting out of line, and the _time_ before you return in to line, is so long, that one who uses the point may easily arrive first.

98

PLATE 28.

This is a hit in _tierce_ on a straight line against a
tierce which has fallen. Both were in _tierce_ on the outside.
You have tried to engage the adversary's sword, being within
wide distance, and he has taken that _time_ and made a cut of
mandiritto at the head. You have withdrawn your body slightly,
have let his sword pass in the air, and merely lowering the
point a little when his sword has passed, have suddenly thrust
in _tierce,_ reaching his body at the moment when his sword
has competely fallen, so that he has been unable to raise his
sword in order to parry or to retreat, or save himself, because his movement was not yet finished, when he was hit.
From this you may deduce the principle that, when possible, it
is always better to let cuts pass without parrying them, so
that you may not be put into subjection and the danger of being deceived while parrying. Further this method is less
fatiguing.

99

PLATE 29.

This hit in _quarte_ against a _tierce_ which has fallen
has arisen as follows: you have tried to engage whilst both
were in _tierce_ on the inside, and your adversary has taken the
time and made a cut in _riverso_ at the head, but being too far
distant he has failed to reach. You realised that his cut
could not reach   and allowed his sword to pass without
parrying, and by lowering your point a little so that his
sword might not touch it, as soon as his sword had passed you
have hit in _quarte_, being defended in that place where the
cut of _riverso_, after falling, generally hits, especially if
it has fallen without effect. Your point has reached his
body at the moment when the _riverso_ has finished its fall, in
such a way that he who has made the cut has been unable to
save himself, because he has been carried forward by his own
blow, and therefore has given you a chance to hit him. If
his arm had been straighter and in better alignment, he could
have defended better. This is the result of the sword missing its object, which, as we have said elsewhere, always
brings more or less disorder. The cut made from the wrist
gives more protection.

100

PLATE 30.

This cut of _mandiritto_ at the head, which is here shown, against a _tierce_ may arise in this manner: you have engaged your adversary's sword, and he has not moved. The swords being engaged on the outside, he has tried to force your sword, and you, feeling the pressure, have yielded, and by bending the wrist, and keeping your hilt close to his sword, have made the cut shown. He has been unable to parry, because, owing to your yielding, his sword has fallen a little, in such a manner that your _forte_,has weighed upon his blade and prevented him from raising it. It may arise also in this way: both being in _tierce_ on the inside, you have tried to engage and your adversary has disengaged in _tierce_, carrying himself forward in order to hit on the outside. But you have let his point drop, bent your wrist, bringing the hilt over his blade, and thus have made the cut of mandiritto. As may be seen your hand has fallen into _tierce._

101

PLATE 31.

This hit in <u>quarte</u> against a <u>seconde</u> has arisen as follows:- both were in <u>tierce</u> on the outside, and you have lowered your point and uncovered the outside in order to give your adversary an opportunity to attempt a hit. He, thinking you have moved in order to change position, has come in to the part uncovered, thinking that he could hit by changing from <u>tierce</u> to <u>seconde</u> so as to exclude your sword on the outside, and carrying his right foot forward. You, seeing him coming, have not parried, but turned the body with the left foot; at the same time disengaging on the inside and changing the hand to <u>quarte</u> you have made the hit. The hit might also have arisen in this way: both being in tierce on the inside, you have tried to engage; he has disengaged before you could find his sword, and changed his hand into <u>seconde</u> in order to cover himself and hit you on the outside in the <u>time</u> when you were seeking his blade. This would have succeeded, if you had tried to parry. But thinking that you could not defend by a parry of the sword, which was moving in order to engage his sword, you have continued the movement followed his disengage, and by making a counter-disengage, turning your body out of presence and letting his sword pass, you have hit at the moment of his advance.

102

PLATE 32.

Now follows another hit in _quarte_ on the outside of the adversary's sword, which is in _tierce_ at an angle. This has arisen as follows: both being on the inside, and the adversary in _tierce_ at an angle, you have tried to engage his sword and he has disengaged in _tierce_ on the outside. In the same _time_ you have changed your hand to _quarte_ without extending the arm, but carrying the hand away to the inside and as high as the shoulder. You have moved the right foot forward turning it in the air in such a manner that the turn is seen to be completed as it reaches the ground, thereby turning the body and bringing out of line all the part opposed to the adversary. You also have made an angle of your sword, which has entered the angle formed by the adversary, as shown. In this way the more the adversary tries to push it away, the stronger is the hit. Or your adversary may have tried to engage in _tierce_ on the outside, your sword being in _seconde_; you have yielded from _seconde_ to _quarte_, turning the body, and hit at the moment when your adversary expected to engage your sword.

103

PLATE 33.

This hit under the sword on the outside may be made in tierce or in quarte against a quarte, according as the arm is parried outwards more or less, and may arise in this way: your adversary has tried to engage your sword on the inside. You have disengaged and he has tried to hit in quarte under the sword. After disengaging you have withdrawn the body, in order to have time to return your sword to the lower lines before he could reach. You have succeeded, and dropping the hand and figure at the same moment have again found his faible with your forte, and hit him in the right side, as he turned. Or you may suppose the position has arisen, when you tried to engage your adversary's sword on the outside, he being in seconde . In that time he has changed from seconde to quarte, turning his left foot, in order to hit under the sword and let your sword pass. At the same moment you have carried your body on to the left foot, returned your sword to the lower lines on the outside, and thus made the hit shown.

PLATE 34.

This *seconde* against a *quarte* has arisen as follows:
your adversary being in *quarte* has tried to engage your sword
in *tierce* on the outside and you have disengaged on the inside,
still in *tierce.* The adversary, taking the *time* of the disengage, has tried to hit in *quarte* in the line uncovered, turning his body. You have changed from *tierce* to *seconde*, dropping your body and sword under his sword and letting it pass
in the air above. Or it might happen that you have moved and
tried to engage his sword on the outside, with the hand in
quarte, in order to have greater strength in the line where
his sword was, and in order to be more covered on the inside.
Your adversary has disengaged on the inside and made a *quarte*
in order to hit above the hilt in the line seen to be uncovered. Then you have changed from *quarte* to *seconde*, and,
lowering your whole body below the position where the hilt
was, have carried forward the right foot in such a manner,
that his sword has passed in the air and you have made the
hit shown.

104

PLATE 35.

Now follows a hit in _prime_ against a _seconde_. Both being in _tierce_ on the inside, you have tried to engage your adversary's sword; he has taken the _time_ when you were trying to subject his sword, and has disengaged on the outside, changing his hand to _seconde_ and advancing to hit over the sword in the line you have uncovered in trying to subject his sword. But you, seeing the disengage and the blow intended, have taken that _time_, changed from _tierce_ to _prime_, lowering the whole body, so that the head is entirely covered and defended by the hilt and right arm, and have pushed out the _seconde_, for in the change to _prime_ your hilt has gone so high as to cover the point aimed at by the adversary with his _seconde_; with the result that his point, which was to hit _above_, has remained _below_ and excluded by your _forte._ Or you may have tried to engage the adversary's sword, and he has tried cut of _riverso_ at the arm in the part seen uncovered; you by a change from _tierce_ to _prime_ have defended yourself and cover- ed your arm with the _forte._ Therefore the adversary has failed to effect anything and has been hit in the same _time._

105

PLATE 36.

This hit in _quarte_ against _seconde_ has arisen in this manner. Both being in _tierce_ on the inside, you have tried to engage your adversary's sword, and he has meant to change from _tierce_ to _seconde_ and drop under your sword in the time, when your point was out of line. Therefore you, seeing his plan, have not completed the engagement, but have directed your point to his body, carrying the hilt where you had planned to put the point; you have turned the body and the right foot, carrying it forward and leaving your hand against the adversary's _faible_. In this manner you are defended and have reached him whilst he was lowering his body and advancing. Equally it might occur that he was in _seconde_ on the inside, and that you have tried to engage his sword. He has intended to disengage in _seconde_ in order to hit on the outside above the sword. You have disengaged, carried the hilt where you meant to put the point, and by the turn of the body, foot and hand, have hit at the moment your adversary thought to hit.

106

PLATE 37.

This hit in _quarte_ against a _seconde_ may arise in two ways: in the first place both combatants might be in _tierce_ on the inside; you have tried to engage, and your adversary has disengaged in _seconde_ over your sword, passing on with his left foot. You, lowering your point without disengaging and letting your arm make an angle to the inside, as is seen, with the hand in a guard of _quarte,_ have turned the body with the left foot, met the adversary as he advanced and hit him in the side under the right arm. Thus his sword has passed idly in the air. In the second place it may be that you have disengaged on the outside, and your adversary has sought to take the _time_ in order to hit above in _seconde._ Then you have simply lowered your point, which had gone to the outside, under his sword, leaving the hand in the same place, but turning it into _quarte;_ without extending the arm you have turned the body and brought all the part which was uncovered when on guard, out of presence.

107

PLATE 38.

Now follows another hit in _quarte_, this time against a _quarte,_ arising in this manner: you have tried to engage your adversary, who was in _tierce_ on the outside. He has planned a cut of _mandiritto in sgalembro_ at the face, keeping his arm in line and working from the wrist only. You have suddenly brought the left foot forward with the point of the foot turned outwards; at the same time you have turned your hand into _quarte;_ extending the arm and bending the body as far as possible, you have met your adversary's sword in its descent, before it was in line, excluded it and hit him in the throat. This is the true method of parrying a cut of _mandiritto_ at the head, when you are forced to parry, for by bringing forward the left foot in this manner, not only does the sword reach further, but it is stronger and can better resist the shock of the cut; with the right foot it is weaker.

108

PLATE 39.

In this case both were in a guard of _tierce_, on the out-side. You have tried to engage by turning the hand into _se-conde._ The adversary has disengaged, turning his body and his hand into _quarte_, in order to hit in that _time_ on the in-side under your hilt. But you have turned at the same moment from _seconde_ to _quarte_ and have brought the left foot forward putting the point of your sword under his hilt, carrying the arm inwards, and the _forte_ towards his _faible_, in such a way that your side is completely defended. It is safer in this case to follow with the right foot, rather than to retire. Such a hit cannot be prevented, even though the swords are of equal strength, because the position of the one who is turn-ing is much weaker than that of the one who is advancing in the manner described; the latter's sword with equal skill will always overcome the sword of the one who is turning.

109

PLATE 40.

This hit in _seconde_ against an opponent in _quarte_ who has
advanced the left foot may easily arise in the following manner: the adversary, being in _quarte_, has tried to engage your
sword, which is in _tierce,_ on the outside. You have disengaged, still in _tierce._ He has attempted a hit in _quarte_ through
your _faible_, advancing the left foot. But in the same _time_
as you disengaged you have dropped your point under his hilt,
also advancing the left foot. By bringing the whole weight of
the body on to the left foot and turning the hand into _seconde_,
you have got far out of the line of your adversary's point and
made the hit. It might arise in another manner: both being
in _tierce_ on the inside, you have moved your point, making a
slight turn of the hand towards _quarte_. The adversary, seeing
the opening, has tried to engage your _faible_ and hit in the
same _time_ by advancing the left foot. But before he has reached your _faible_ you have dropped your point under his hilt, so
that he has failed to find your point, and in the same _time_
have carried your body out of line, bringing the weight on to
the left foot, which has advanced. In this low position you
have been able to penetrate to his body, as you were already
well advanced. Or again, both being in _tierce_ on the outside,
the adversary has tried to engage your sword; in the same
time you have threatened a cut of _mandiritto_ at the head, us-

ing the wrist and keeping the arm steady. He has changed from _tierce_ to _quarte_ in order to defend the head, and advanced the left foot in order to hit in the same _time._ At that moment you have checked your sword near the adversary's, without touching it, and immediately changed your hand to _seconde_, lowering the point under his hilt, advancing the left foot, with the body so bent, that his point, which would have hit in the chest, has passed over. Therefore you may see how dangerous it is to parry, even with a thrust in the same _time._ Therefore, unless forced, it is always best not to parry.

111

PLATE 41.

Here is another hit in seconde also against a quarte. Both were in tierce on the inside. You were in a stronger position than your adversary and have made a feint of hitting in quarte through his faible. He, thinking the thrust was coming, has made a turn of his body with his right foot and a thrust in quarte through your faible, in order to meet you in the time of your approach. Seeing his plan, you have suddenly changed to seconde, lowering your point and body and bringing the left foot forward; thus you have made the hit by continuing on to his body, before he could recover, for he has not passed, but turned, and his left foot has remained . steady. Or it may be that you have tried to engage your adversary's sword on the outside. He has disengaged in tierce on the inside, but in that time you have made a feint in quarte. He has tried a counter quarte through your faible turning his body out of line, in order to meet your approach. Seeing the danger you have changed from quarte to seconde and made the hit shown, while his sword has passed over in vain.

112

Pl. 42.

Another hit in _quarte_ against a _tierce_. You were in _tierce_ on the outside, as was your adversary. You have made a feint of hitting in this tierce on the outside, and he has moved to parry and hit by pushing on his right foot, enticed by seeing you move without a _time_. Seeing your adversary moving to parry and hit, you have placed your left hand on the inside of his sword, disengaged in _quarte_, advanced the left foot and so hit him at the base of the right side. Or you may have been on the inside and may have disengaged with a feint of hitting on the outside. Your adversary has tried to parry and you have placed your left hand on his sword and made the hit. These defences with the left hand are here shown in order to demonstrate how in case of necessity only, they may sometimes be used. The effect is seen, and you may realise how easily such defences may be deceived. Towards the end of the book we shall describe a method against which the left hand will not prevail nor parry.

113

Pl. 43.

This next _tierce_ against a _quarte_ has followed when both were in _tierce_ on the inside. You have made a feint of hitting in _quarte_ on the inside. Your adversary has tried to hit in counter _quarte_ through your _faible_. In the same _time_ you have lowered your sword hand to tierce, carried your left hand to his approaching sword, lowering and turning the body with the left side forward, so that your hand has carried his sword away and you have hit him in the chest. It might arise in another way: you being on the outside have pushed the adversary's sword away. He has tried to disengage and hit in _quarte_ on the inside. You have parried with the hand and hit him below as shown. Or it might very well be that both were on the outside, The adversary has tried to engage; you have changed your hand to _quarte_ in order to avoid the engagement; he has tried to hit with another quarte in the line seen to be uncovered, and in that _time_ you have parried and made the hit.

114

Pl. 44.

The next is a hit in _seconde_ against a tierce. Both being in _tierce_ on the outside you have made an _appel_ by turning the sword from tierce to second and carrying the point inwards out of line.You have brought your left side so far forward as to uncover the whole chest to the adversary, but with the sword so low that he could only hit above, and holding the left hand before the face. While your adversary has seized the _time_ to hit in the part uncovered, with the left hand you have pushed his sword outside your left flank, in the same _time_ advancing the left foot, and with the body low have disengaged in _seconde_. Thus you have made a hit in the chest by extending the right arm as far as possible and bringing forward the right side also, but with the point of the left foot turned outwards in order to carry the body away from his sword. The result is here seen.

115

Pl. 45.

This is another seconde, but against a quarte, with the
right foot advanced.   Both combatants being in tierce on the
outside, you have disengaged without waiting for a time or provocation of your adversary. He has seized the opportunity and
tried to hit in quarte. You have at once turned your hand into seconde, brought the left side of the body forward, turned
the heel of the right foot, placed the edge of your left hand
over his sword, and hit in seconde in the chest. It might
have arisen from both being in tierce on the inside; you have
lowered your sword, leaving yourself uncovered, and he has
thrust in quarte. Then you have raised your hand into seconde,
changing the front of your body and keeping the right side
back, as being in the most danger. In this manner you have
parried with your hand, for this low quarte is forced down by
the parrying hand whereas the point would naturally make a
hit in the chest.

116

Pl. 46.

This is a _quarte_ with a turn, which has hit against another _quarte_ with the left foot advanced. The one who has passed has made a feint of hitting on the outside over the sword, and you have moved to parry. The adversary has placed his left hand on your sword in order to parry, and in the same _time_ has disengaged in _quarte_ on the inside, advancing the left foot, so as to hit in this _quarte_. But you, who have moved to parry the feint on the outside, seeing that your adversary was going to defend with the left hand, have disengaged your sword, which was above, on the outside of his hand, and thrust at his advancing body, bringing yourself out of line with a turn of the left foot. You would not have hit so low with the point, had you not found his _faible_ with your _forte_, so that you were more defended. In this manner the attempt of the hand to parry has been deceived, as shown.

117

<u>Pl. 47</u>.

This is another <u>quarte</u> hitting against a <u>tierce</u> designed to hit under the sword. You have made a feint of hitting towards the right side of your adversary's face. He has tried to parry with the left hand, lowering his body so as to hit under the sword on the inside. But you, who have made a feint, have seized the <u>time</u> of his raising his hand to defend the head, lowered your point to the space between his two arms in the <u>time</u> of his making the opening, and, changing your hand into <u>quarte</u> and turning the body with the left foot, have made the hit. The adversary has been unable to parry, because your sword was shut in between his two arms and could not be pushed aside without a change of plan.

118

Pl. 48.

This is the last hit, in _quarte_, against a _seconde._ The adversary meant to parry with the left hand but has failed. Both were in _tierce_ on the inside, and he, who has tried to parry, has so far, withdrawn his guard that his _forte_ could not defend him and he has trusted to the defence of the hand only, which was too high for the face. You have made a feint of hitting in the angle of the right side. Your adversary has turned his body in order to withdraw that part, carried his hand to the defence and changed to a guard of _seconde_, in order to make a hit in the chest. Seeing his purpose you have disengaged your sword from the line of the fingers of his hand and hit him in the chest in _quarte_ in the _time_ of his advance. Turning the body out of line you have also covered yourself with the hilt, so that his sword has passed in vain, although the angle of his _seconde_ was directed towards the line into which you were turning the body.

119

SECOND PART.

DISCOURSE ON THE PRINCIPLES OF THE SWORD AND DAGGER.

As we consider that we have treated at sufficient length of the practice and times of the sword alone, it has seemed to us suitable, in order to give full satisfaction to the reader, to introduce here the instructions and rules of the sword and dagger.   We do not mean to prefer this method of arms to that of the sword alone, nor to multiply different precepts, nor to throw over the instructions given by us elsewhere.   It is rather our purpose to show the wealth and abundance of the practice of this art, which by uniting several arms together becomes more admirable and perfect.   Therefore without neglecting the subtleties of the times of the sword alone, or denying them in any way, for they are beyond comparison the more artful, we shall now leave them, having put them forward and analysed them in their place, and shall proceed to describe, as far as is necessary, the perfect use of the sword and dagger, and to impart the true knowledge of the stratagems useful in the attack and the defence, and dangerous to those inexperienced in the art.   One who is well acquainted with the times, will easily find great benefit.   These two arms are allies of one another, and by their union give great strength in need;   also they divide and share the functions, the one defending and the other attacking.   We hope those who practise with these arms, following the present instructions, will reach the perfection they desire.

120

THE POSITIONS OF THE SWORD AND DAGGER.

The positions of the sword and dagger are formed with the body bent and the feet close together, and with the weight of the body on the foot which is to remain steady. The arm holding the dagger should be extended as high as the point of the shoulder, with the point of the dagger inclined upwards and directed towards the adversary. The sword should be so far advanced as to extend beyond the dagger by at least one fifth of the sword's length, so as not to be impeded by the movements and feints of the adversary. With this guard you can make better use of the _forte_ of the sword in any event. The sword and the dagger should be in conjunction so as to close the path between them against the adversary. If you wish to use the sword in an advanced position, the dagger should be held in conjunction with the hilt of the sword, so as to close the path to the adversary. Similarly the right side should be kept forward and the left back, as the latter is more exposed and in greater danger.

121

THE COUNTER-POSITIONS.

The counter-positions are more difficult to form, since you must attend to your own two weapons and the two of your adversary and their positions. You must be careful not to advance so far in the desire to acquire some advantage, that your adversary can engage your sword with his dagger and hit. Therefore you must keep the sword so far distant that you know you can save and move it before it is engaged by his dagger. For the rest you should observe the same conditions and principles which we explained in describing the counter-positions of the sword alone, taking care that your body is clear of the adversary's sword without movement of the body or weapons, and that your weapons are in conjunction to give greater strength and surer defence.

122

ON ENGAGING THE SWORD.

Engaging the sword with the sword and dagger is very different from engaging with the sword alone, for it is done at one time with the dagger, at another with the sword, and more often with both sword and dagger. You engage with the sword when your adversary keeps his sword so far withdrawn that you cannot reach it with the dagger; sometimes your sword cannot penetrate because of the danger of his dagger. In this case you should hold your sword in such a way as to close the path in which the adversary's point is directed and with the forte in such a position that his point cannot approach your body without his _faible_ meeting your _forte_, which generally prevents him and drives his point out of line. You can also, as he approaches, raise the sword from that defence and make a hit, putting the dagger where the sword was before, which will be a good defence and will delude your adversary, who will probably think that you are going to parry with the sword; thus his plan will be foiled. This is a good method to follow, when your adversary's sword is withdrawn, since you should not advance too far with the sword, lest you lose it, nor should you advance the dagger, for there are some who, seeing the dagger advance, beat it with their dagger and hit. Therefore to advance too far is very dangerous. Besides when you push on so far, the dagger is easily deceived by the feints and movements of the adversary's sword, which throw it into

disorder. But if your adversary's sword la advanced, there will
be no such danger. You can then readily try to engage with the
dagger, if you understand the correct method. For you must not
carry the dagger so high that on reaching his sword you have to
lower it, or so low that you have to raise it; nor on reaching
his point must you make any movement for, however slight, it
would give him an opportunity to hit, or at least to disturb
the dagger; thus you would not be free from danger. You must
hold the dagger with the point in the same line as your adversary's sword, so that on reaching his point the dagger engages
without other movement. If the line of his sword is rather
low you must begin with the dagger equally low. To make safe,
the body should be lowered in proportion and in such a manner
that you know that if your adversary disengages, you can easily
parry without raising the arm, for if you raise it, he might deceive you with a feint of disengaging and return,when you
would be hit without a defence. But holding the dagger arm
steady, you would easily defend, on either side. Thus, if
you know how to apply the exact rule, you can engage your adversary's sword, wherever it is, provided it is far enough
advanced beyond his dagger. But if it is withdrawn behind the
dagger, it would be an error to engage; for besides the reasons given you would be in danger of getting within too close
distance before finding his point, when you would be hit, as
you would also if his point were too low towards the ground.
In that case it would be better to cover it with your sword,

so that your adversary could not disengage on the side of the
dagger, or if he wished to disengage would be forced to disengage on the side of the sword in order to free his point.

But if you wished to hit, it would be necessary in forming this defence, to hold the dagger steady; thus you could
turn the sword from _tierce_ to _seconde_ and close the path between the two weapons in such a manner that, if your adversary
advanced, his sword would always encounter your dagger, which
had made no other movement. But if his sword were low and
on the outside on the side of the dagger, you could not cover
the path, and it would be ill-judged to advance the point of
your sword because of the danger, lest in that time he should
disengage above the sword and hit between the two hands where
the path was open; although you might parry with the dagger,
yet the movement would be so large, that, if your adversary
had made a feint of hitting in that part he could hit in the
new opening and attack. Also he could disengage the point of
the sword, thrust between the two hands, gliding along your
blade, and thus hit on the side of the dagger, so that you
would be in the same danger. Therefore in such a case it
would be better to push his sword out with the hand as low as
the right knee and turn   to _quarte_, in order to be on the inside and cover that part more; at the same time you should incline your point upwards towards the dagger, which would defend
all that part of the body up to the head, and all the more if
dagger and sword were in conjunction. In this manner you would

not only push away his sword, but would form a good counter-
position. You would have left the adversary only one path of
attack, that is over the sword, and your sword being at an angle
with the left arm forward, you could easily parry this attack
with one of the two weapons or both together and with little
movement, since there would be no room to hit between the two
weapons. Whether your adversary's sword were in <u>seconde</u> or in
<u>prime</u>, and if his point was in that same line but separated
from the dagger, which would mean danger below and between the
weapons, it would still be better to use this method, and to
put the point of your sword against his, which would lead to
the same result. But if his point were in another line and
high, in that case it would be well to adopt a guard in <u>tierce</u>
at an angle, and put your point on your adversary's, holding
the dagger extended near the sword;in order to prevent his
hitting anywhere but above the dagger. For if he were on the
outside of the sword and should thrust over the dagger, he
would make such a large circle that he would give you a good
chance to parry, if you remembered to keep the lower part of
the body so far withdrawn that your adversary could not reach
it. In any case you must when parrying hit where there is a
part uncovered according to the movement and position of your
adversary's body and weapons in making his hit. As he have
many times said in this way the parry is always safer and cannot be deceived.

126

You must know that the principle of using the dagger alone in engaging the adversary's sword applies more against the guards in <u>tierce</u> and <u>quarte</u>, than against those of <u>prime</u> and <u>seconde</u>, where it is not so strong. For if when you move with the dagger against your adversary, he should make a feint in the upper lines, which are the feints most to be feared, he would force you to parry and then would hit by a rush in the parts uncovered by your movement of parrying. Besides the fact that they are more successful with the sword and dagger than with the sword alone, the guards in <u>prime</u> and <u>seconde</u> naturally lend themselves to the rush. For being in <u>tierce</u> or <u>quarte</u>, in order that the rush may have force, you must change the hand to <u>seconde</u>.   Therefore when your adversary is already in a position to rush without further change of the hand, he is much more to be feared.

If then you wish to guard against your adversary's attack, it is better on his advance to use the sword and dagger together and for greater precaution to exclude his point on the outside, if possible. This method is good against a guard of <u>prime</u> or <u>seconde,</u> though you must take care if you wish to advance with safety, that you have excluded his point entirely in such a way, that you are certain he cannot hit there, but without touching his sword. If you have done this, you may close the distance as much as you wish. You must also be careful not to approach the point of your sword so near to his dagger, that he can engage it and hit before you have freed it.

127

You must always keep your point at such distance and position that you are assured of its freedom to hit in _time_. Nor by freedom do we mean keeping it far out of line for in that case it would be pushed away before you could return it. But holding it in the proper manner you will keep it in line or very little out of it, and always free, so that your adversary cannot prevent your returning.

We must add that you can engage your adversary's point in any position with the sword alone, provided that you hold the dagger in such a manner that it has little movement to make in order to defend the part where the adversary might hit. This is an excellent method though some deny it. They will not allow that you should ever engage with both the weapons at one time, but they say that one at least should be free, in order to be able to parry and hit if the need arises, that one weapon should be reserved for defence, and one for attack and that there being two pieces they should serve two purposes, whereas if both are used for the defence they are serving one only. We admit this, but say that this united defence is not only stronger, but also better protects the other line, where the adversary might approach. He finds little exposed there, it is harder for him to hit and easier for you to parry. Further if you defend with one weapon only, there is more danger not only of being disordered, but also of being overcome. Thus it often happens, that where you are defending with one weapon and your adversary changes his line in hitting, you are so disunited and

weak that both your weapons are forced into subjection, so that what you would not do willingly, you are forced to do, when your adversary has moved. Thus you are so disordered and confused that you have been unable to hit because of the trouble you were in in the defence. On the other hand when you defend with both weapons and your adversary changes his line in order to hit or do anything else, you can on that change separate the two weapons, the one to parry and the other to hit because they were in union. Sometimes also you can defend with both weapons and hit in the same _time_ because of the strength of your first defence due to the union of the weapons. You have more completely covered the body and work of your own accord and not driven by necessity. Therefore you may well understand, that he who engages with both his weapons will dispose of them with greater judgment and security and in such a manner that he will not be prevented from hitting in _time_,when the occasion offers; but he who is compelled to engage will usually be prevented from making anything but the simple defence, and however good that is it can easily be deceived.

129

HOW TO PROCEED AGAINST AN ADVERSARY ON GUARD ON THE LEFT FOOT.

In dealing with an adversary who is supported on his left foot, you must consider that his sword is so far withdrawn, that it is difficult to engage it, and that he holds his dagger well advanced with the idea of engaging your sword and then making a hit in <u>time</u>, generally with a pass; for he realises that his line is short and that he cannot reach without a pass, and he is well aware that he cannot pass without first engaging your sword. You must consider too that by passing he will come with a great impact owing to the great distance from the spot where he lifts his foot to the spot where he brings it to the ground, and owing to the fact that his sword is carried on not only by the arm and foot, but by the whole body. He advances his body with great vigour in order to make his hit quickly, and therefore he strikes violently, especially if he comes in the <u>time</u> of your advance, when you cannot break ground, and thus the encounter is all the more impetuous. Further, since your adversary recognises that he can do nothing whilst he is within wide distance, he will always seek to approach in order to gain your sword and the required distance.

From all these considerations you should be able to dis-cern your advantage and avail yourself of it, that is to say the distance, in which you may arrive, more quickly than he who is supported on the left foot even with his right foot in front. While he is seeking to engage your sword is the time to hit and

to break ground so as to keep away and prevent his passing, or, if you cannot hit, at least to make a feint in order to disorder him and then hit, or play him and in that _time_ withdraw so far, that he will remain at the original distance. Then you may choose a more convenient _time_,when he moves again, for since he is on the left foot,his side below the dagger is in great danger, and if he tries to cover it, he uncovers himself above, since the dagger cannot cover both places at once. Therefore choosing the _time_ you can always hit in one of these two places. Everytime that your sword provokes his dagger to move you will certainly hit in one part or the other, and all the better if the provocation is in the _time_ of his advance, when he cannot break ground. Also you must not advance so far that you are in danger of receiving a _riposte_ stronger than your thrust, as often happens. When you see the adversary advancing in order to engage your sword with his dagger, then you should hold your point in line with his fist; if his dagger is extended with the point forward so as to hide his fist, then you should hold the sword straight under the blade neither inside nor outside, and therefore you must hold it in a straight line in _tierce_. As your adversary approaches, you must bring the arm back to the body keeping the point in the straight line and not letting it drop, and draw him on until his hand penetrates the point of your sword; at the moment when his foot arrives within distance, then hit in the straight line under the blade of his dagger, the nearer the arm the better.   If

that part is not uncovered then you must hit over the dagger,
making a slight turn of the hand towards seconde, but close to
the dagger, above all taking care to arrive quickly. If the
part uncovered is on the inside, you could turn the hand to
quarte so as to hit that part, but still close to the blade of
the dagger. These thrusts will certainly succeed, if you take
the time of your adversary's advance. If his dagger is held
so exactly that you cannot hit, you should move the point a
little inwards or outwards in order to make him waver and then
hit. To protect yourself better against the riposte or counter-
time which your adversary might make,you should hold your dagger in such a position towards the point of his sword, that,
wherever it comes, the dagger can parry with little movement.
In this position you may be certain of having greater ease in
defence; but you should not be already so far advanced that
your adversary's sword can harass you, before you are at the
proper distance.

If your position is not so subtlely formed, and the adversary's dagger begins to penetrate your point on the one side
or the other, you should not therefore disengage with the idea
of freeing it, bat should gradually bring it out of line as far
as his dagger can follow it. For if he tries to engage it, his
dagger will go so far out of line that he can be hit, or if you
make a feint he will be so disordered by the large part uncovered, that he will then certainly be hit.

If you do not wish to follow this method, you should remain

within wide distance, with your sword free and somewhat drawn back, so that it may not easily be engaged by your adversary; you should incite him by various _times_ and _appels_, being always ready to break ground, so that when he decides to hit and seizes a _time_ to make a determined pass, you can protect yourself according to the quality of the guard. For if you go to meet him although you may protect yourself from the first effect, still the parry will be so violent, that your power to hit will be taken away. Moreover if your adversary changes his line, you will certainly be hit. On the other hand if you break ground your adversary's thrust will lose its force and may easily be parried; even if your adversary changes his line, you may still parry in the new line before he arrives. Therefore one who stands on guard on his right foot should not try to get in this close distance against an adversary who stands on the left foot but should keep at a distance; he should advance in order to strike when the adversary approaches, making use of the advantage of his longer reach.

There is still another method, when you intend to defend and break ground, that is to place your sword on the adversary's dagger, so that when he thinks he has engaged your sword in the _time_ of his advancing you may free it by breaking ground and hit. This is a subtle method and deceptive when used with the necessary circumspection,so that you are not deceived in thinking to free your sword and hit. You must take care too that your adversary is not aware of your intention, and may not make a mock of it.

133

WHEN ON GUARD ON THE LEFT FOOT HOW TO PROCEED AGAINST AN ADVERSARY ON THE RIGHT FOOT.

If you wish to advance with the left foot first in assaulting an adversary on the right foot you must realise your disadvantage, that is to say, your sword in hitting and recovering does not reach so far as your adversary's, who is on his right foot and may hit and easily recover with little movement of the body. Therefore you must bring the right foot forward and when you have passed, owing to the great movement of the foot and the distance you have advanced, it is impossible to recover and break distance in one <u>time</u>. Therefore it will be necessary if you wish to recover after hitting to weigh on your adversary's sword and hold it in subjection towards the ground until you have recovered, in order to prevent his hitting while you are withdrawing. Even so, if your adversary were skilful in freeing his sword, you would be in danger of being hit before you had come to rest owing to the great distance the foot has to recover, even longer than its advance.

Therefore when you have advanced the right foot it is better to follow with the left also, and pass entirely, or having advanced the right foot and hit, at once to recover your weapons to the line of your adversary's sword and close its path entirely, remaining steady on the feet. If your adversary withdraws in order to free himself you can hit again in the <u>time</u> of his movement for within such close distance he cannot get back in

time. All these rules apply after you have made a hit. In order
to approach within striking distance, you should, for greater
security, hold your dagger in such a manner, that as you advance
the foot it reaches the point of the adversary's sword near to
the blade without any other movement, so that you are sure that
the dagger in that position defends the straight line from the
point to your body. This is the true method of acquiring distance, so that the adversary cannot easily hit; if he moves his
sword, it will be a change made to avoid the danger, which will
give you a chance to hit as he moves. If you are not within
this distance, you must approach with little movement of the
point of the dagger and cover his new line in the same way.
In order to find a chance to hit and pass, you should glide
along the blade of your adversary's sword with your dagger,with-
out beating it, that is to say when you are on the inside. If
on the outside you should neither beat nor glide along his blade
but leave the dagger in its place, or little in advance, since
you must never penetrate the fourth part of his sword. In hitting you must turn the hand into the guard of quarte in order to
unite it with the dagger, thus defending the upper and lower
parts at the same time.

It is true that sometimes this method with the dagger is
impossible, because the adversary holds his weapons so close
together or because of the angle of his sword. The dagger is
not sufficient to defend from the knee to the head, for it cannot defend more than one place at once, and in advancing with-

in distance it might be harassed and disordered so much that the adversary by observing the _time_ and taking advantage of the movement might forthwith hit. Therefore, as you have your left side forward and no other defence is possible, you must add the union of the sword, that it may defend one part, while the dagger defends the other. In this way there will be great benefit to your body, for the dagger will be more secure in its defence, and you will have greater advantage in hitting, since the point of your sword will be always nearer to the adversary and no less safe from his weapons than before. Now we will show the position in which the sword should be held.

You must be in a guard of _tierce_ with the point directed towards the point of your dagger, with the hand so far advanced that you know the _forte_ can defend the side under the dagger and with the dagger arm so far extended to the outside that you know too that there is nothing uncovered over that arm. The dagger must be accompanied by the sword, so that the part below is defeated when on guard, and your adversary can hit only over the sword on the outside. The points must be so close together that the sword cannot be separated from the dagger and you become in danger of a hit in that weak place. In brief the point of the sword must be fortified by the dagger, so that the adversary's sword cannot thrust it away and hit. This is the position of the guard with which you must advance within distance in order to hit when an opportunity offers. In approaching you must protect the part uncovered over the sword by keeping the feet al-

ways outside the adversary's sword and the body sideways with
the left side forward but bent, with the head over the hilt of
the sword so that if the adversary approaches you can more
easily defend that uncovered part. If you bend the body in the
natural way and hold the sword so as to defend the left side,
the head will be so far out of the line that it would be hit
before it could be defended. But if you bend it over the hilt
and the left knee in the manner described, with the shoulder
turned to the inside, you will so place your body that none of
it will be outside your sword; therefore if your adversary
wished to hit, he would necessarily have to hit close to the
blade of your sword, and thus the defence would be easy. In
this position you can advance against all guards of the right
foot, high or low, with the sword straight or at an angle. The
only change necessary is to hold the points higher or lower,
more to the inside or the outside, according to the position
of the adversary's sword; but his sword is on the outside and
very high, you must change your hand from tierce to quarte still
observing the conjunction of weapons in order to defend the
angle of the guard in seconde, so that he may not have an opening to place his point under the dagger. If his sword is on
the inside you must observe the same union, but it would be better to have the hand turned rather more to quarte than to tierce
in order to be stronger in that line. If your adversary disengages, you must hit without any other movement of defence by
merely extending the sword far enough to reach.

137

GENERAL DISCOURSE ON THE USE OF THE DAGGER.

In practising we say that it is good to hold the dagger with the arm extended with little help from the sword, so that you may become secure in the defence and may make little movement in parrying in order not to uncover one side while covering the other; for when the dagger is held well forward and you are disordered by large movements, you become confused and are defeated. Therefore it is necessary to use it judiciously and with practice to acquire such exactness that the dagger is not disturbed by the movements of the sword and you are more secure in its use. When then you have acquired that familiarity and security, you can hold it withdrawn and push it forward to the defence as required. Then you will be certain that it will perform its function with more exactness and security in engaging the adversary's sword. Similarly after such exercise you will be able to hold it more in union with the sword, which will render the defence easier and stronger, nor will your dagger be so much harassed. But you should not hold it fixed in any position. The reason is that in some cases it is good to have it advanced, and in other cases not according to the position of your adversary's weapons, whether advanced or withdrawn, and according to the position of your own guard. For having fixed the body and pushed the sword forward, it may be better to have the dagger withdrawn, while in other cases it may be better to have it advanced. But to discuss all the positions in which it may be

held against the various counter-positions would make this discourse too long, for the subject is very large, almost endless. We shall merely say that as a general rule the point of the dagger should be held always opposed to the point of the adversary's sword, until that point is turned against the body. To hold the dagger out of line would not be advantageous, nor to keep it turned towards the feet, unless you were careful to hold the body so low that you could defend the upper part with little movement, and were ready to hit in the same _time_. As we have said it is better to cover the dagger with the sword in such a manner that your adversary's only resource is to disengage. This method is safer and less subtle, and may also be used against a sword held high. Also you should hold the dagger so close to the sword that there is no path for the adversary's sword between the two hands; his sword will either meet the dagger or pass on to the _forte_ of your sword. We must also remind you that in using the dagger the defences are all to be made with the edge, whether on the inside or the outside, whether high or low, and with the strongest part of the dagger especially against cuts, and against thrusts too, when you wish to glide along the blade. It is true that if you wish to beat a practice which we do not recommend, it is better to do so, with the point of the dagger on the _faible_ of the adversary's sword, since the point describes a larger circle.

Pl. 49.

139

This discourse will explain the guard in <u>prime</u> with the sword and dagger here illustrated. The hand is in the position reached by drawing the sword from the scabbard. With these weapons your guard will be securer than with the sword alone, for the dagger defends the upper part towards the face, which is nearest to the adversary and defends below to the middle of the body. Below that part there is no danger since the adversary cannot reach. If you keep the feet close together and the weapons in conjunction, your sword cannot easily be engaged and is always free, whilst the conjunction of the dagger with it prevents the entrance of thrust or cut between them. Cuts at the head too are defended by this guard, and you may hit in the same <u>time</u>. With this guard in every case after hitting you should recover to the same position for your protection, keeping the dagger always extended. As to the legs no other defence is needed, where the adversary cannot reach. In the <u>time</u> of your adversary's approaching to hit you can thrust a cut at the head or sword arm, merely bringing back the left foot, thus you will protect yourself and make a hit.

140

Pl. 50.

The second guard in order is a guard in _prime_ in its nature and derived from the first. The difference between them is that you have carried the left foot forward, or the right foot forward, and that is the only change in position. As to its security this guard is inferior to the first, because the advanced leg is in great danger and the side below the dagger is exposed. Nor can you hit without passing, and having passed you cannot return on guard without being hit owing to the length of the movement; also you are too far advanced to be able to break ground. Still this guard may be used, if its correct principles are observed, that is to wait for the adversary to hit him in order to parry and hit, or make a feint with the point and cut. A guard formed in this manner is also fitted for making a feint with the point and thrusting; but you must not give the adversary a _time_ or an opening, unless you carry thw weight of the body on to the rear leg, leaving the other leg exposed; when the adversary advances to hit that leg, you must carry it behind the other, which you may do without difficulty as it is already relieved of the weight. In this manner with the body bent forward on the right foot you may in the same _time_ thrust or cut according to the opportunity; this is beyond comparison the best device which can be used with this guard.

141

<u>Pl. 51.</u>

This plate represents an extension made from the guard in <u>prime</u> with the sword and dagger. This guard would naturally hit over the adversary's weapons, for if you try to hit below with this extension your adversary would easily knock your sword to the ground. The plate shows the dagger extended and the body bent in order to parry the blow which the adversary might make; for in hitting you must never withdraw the dagger or let it fall-back, since it is clear that in the <u>time</u> of withdrawing or abandoning it you cannot parry. With this guard you increase your danger by the large part uncovered by the angle formed with the sword hand, so that if you could not parry you would be hit,and still more easily as the thrust with this guard is shorter than with the other guards. After hitting with this thrust you must recover the right foot to the other foot and wait for a new opportunity; if your adversary does nothing, you can make a feint of a thrust on the inside of the dagger, but if he should design to hit in <u>tierce</u>, as he may easily do, you should then make a cut of <u>mandiritto tondo</u> under his dagger, which will hit him in the sword arm at the moment of his thrust in <u>tierce</u>; you must carry the right foot forward a very little way according to his distance, parrying his point at the same moment with your dagger by pushing it out to the left side and bringing the right side forward in order to facilitate the defence and lengthen your sword; this will have a good effect when you recover on guard.

142

Pl. 52.

The next is a guard in <u>seconde</u> with the sword and dagger, better and more convenient than the guard in <u>prime</u>, because the arm is not so strained, and safer, because the sword covers the lower part and keeps the adversary at a distance. Further cuts at the head can be parried with this guard, but with the weapons in conjunction for greater strength, and in the same <u>time</u> you may hit. Thrusts are parried with the dagger alone. This guard will be most successful if you are careful to hit with the right foot, raising the left foot and carrying it somewhat back, but in a circle, and recovering the right foot close to it. In hitting the right side should go forward. Carrying away the foot must be done with great care, so that the body in recovering may describe a circle and get out of the line of the adversary's point, while you are covered from the line of the dagger, without movement and the distance is enlarged. In closing distance also you should move in a circle towards the adversary's right side and with short steps,in order not to close except on hitting; also you should keep the body out of distance while continuing to move in a circle. When you hit,advance in a straight line and recover in a circle. With this caution you will be very secure in this guard in comparison with some others. With the same short steps as in approaching you can also withdraw, preserving the guard without any disadvantage, or you can proceed according to the occasion by advancing more or less without any change of line.

143

Pl. 53.

The position here shown arises after you have carried the left foot back, leaving the sword and the right side of the body advanced; thus the dagger has been shortened and remains near the hilt of the sword; the left side too has been left exposed above, but at such a distance that the adversary cannot easily reach it. It is seen that you are carrying your feet towards the adversary's left side, in order to withdraw your own left side and find an opening to hit him above or below the dagger by advancing the right foot in a circle towards that part with the same motion of the body, and without moving the dagger from its position. In this way if it happens that your adversary tries to hit the exposed left side by closing the distance, you could while parrying change from seconde to quarte and hit at the same moment, and then recover in a circle to the same guard. This is the true reason of this position; he who knows how to use it can also provoke his adversary to hit by a time or an appel, in order to use the counter-time. Moreover the dagger is safer in this position than with the arm extended.

Pl. 54.

144

Now follows a guard in <u>seconde</u> formed with the left foot,
much better than the first one formed on that foot, becausethe
body is more ready for every event, and more protected by its
low position so that you can both hit and parry with less movement. With this guard you can close distance on the outside of
the adversary's sword, while engaging his sword with your dagger. With this guard too you can hit by advancing the right
foot, always leaving your dagger on his sword without beating
it. You can pass right on to the adversary's body; but if
you wish to recover you must meet his sword with yours and disorder it, so that he cannot hit while you are recovering, since
you will actually have passed so far forward that you cannot
break ground in one step. If your adversary's sword is on the
outside of the dagger and he offers a <u>time</u> to hit, then you must
change from <u>seconde</u> to <u>quarte</u> with the sword and dagger in conjunction, so that you may be defended above and below the dagger,
and turn the body in order to lengthen your thrust and be better
defended; in order to recover safely you must let your sword
fall on the adversary's in a guard of <u>tierce</u>, after hitting.
As to engaging his sword without being deceived by a disengagement, you must remember when within reach of his sword with your
dagger to put him into subjection at that moment, by showing that
you intend to hit, so that by that fear he may be prevented from
freeing his sword, except by fleeing from the danger and retiring.
In that case you should not advance nor do anything but try to

engage his sword again without beating it, so that he cannot hit in any lime, you must then take care not to let your dagger fall in order to avoid the danger of being hit on the outside over the dagger.

Pl. 55.

This next guard in seconde with the right foot forward, the point of the foot turned outwards and the whole weight of the body on that foot, is really an excellent position, because the exposed parts of the body are well withdrawn and in consequence there is little the adversary can reach. Moreover in this position the body can pass with great swiftness. In advancing take short steps and try to engage your adversary's sword; when you have found it with the dagger, do not beat it but glide along the blade, and hit by passing in order better to change the front of the body and to bring out of the line of his point all that part which might be hit. With this guard you may hit in quarte as well as in seconde. Also you give the adversary no chance of a time but press him resolutely and assault him without changing time or being disconcerted in any way.

146

Pl. 56.

The guard shown in the next plate is also a guard in seconde, but little used and perhaps unknown.   The sword is held
foreshortened, the dagger upright, the body bent, and the weapons
so low that only the part on the outisde over the sword is exposed, and there only can the adversary hit.   This guard must
be carefully formed, with the sword somewhat lower than the adversary's as long as his point is as high as his hand.   If his
point is lower than the hand, you must proceed above the part
covered, taking natural steps and moving in a circle towards
his right side;   by this means you will withdraw the part above
the sword which is exposed and, when you have found his sword
with your dagger, your whole body will be covered when on guard.
You should never proceed, until you have first advanced far
enough to engage the adversary's sword with your dagger;   but
when engaged, you should continue resolutely with natural steps.
The lower the body is the more successful the stroke will be.
You must make the hit with the body low, without stopping or
waiting, but always continuing in a circle.   Even if your adversary should thrust, you should not stop but keep the union
of your weapons;   you must not become disordered nor rush, but
glide along the blade of his sword with your dagger as far as
the hilt.   Thus you may hit in _seconde_ or _tierce_ or _quarte_ according to the opportunity.   But throughout you must always
keep the weapons in conjunction both when on guard, and when
hitting in whatever manner, so that the adversary cannot enter
with his sword between your hands.

147

<u>Pl. 57</u>.

Now follows another guard in <u>seconde</u>, with the feet level and separated, the body bent forward, the chest directly facing the adversary and the arms and weapons curved and so high as to cover the whole head, which cannot be attacked by the adversary except below and between the weapons. Since the chest faces the adversary's point when he hits you can move with whichever foot you please, carrying it into the line of the other; in this manner your whole body will be brought out of line, for if his point is exactly in the middle, it must also be in the middle of the two feet, and when one of them moves, his point is necessarily out of line an amount equal to the half of the step you have made. Thus you can in the <u>time</u> of his movement hit, pushing his sword outwards so that it cannot return into line. As both feet are in line with this guard your reach is long and you may go right to his body and recover by carrying to one side the foot which has remained steady, and again turn your chest facing his sword. With this guard you may wait or attack the adversary without his giving a <u>time</u>. It is very secure against cuts as the head is defended on both sides, so that it is not necessary to move the weapons to defend it, while below the legs are so far distant that they cannot be reached. We shall speak of his guard again in treating of the attack with resolution.

148

Pl. 58.

Here follows an extension from the guard in <u>seconde</u>, differing little from the extension from <u>prime</u>. The difference is in the sword hand, which is turned slightly upwards; therefore the extension is longer and the sword hand nearer the dagger, which gives greater security between the weapons. With this <u>seconde</u> you can hit in any part, so that it is better and more convenient than the <u>prime</u>. You may use it on more occasions and <u>times</u>, whilst the union with the dagger offers greater protection.After asking a hit you must be careful to withdraw the right loot close to the left and recover to the same guard. In closing distance you must make a circle towards your adversary's right side and hit by passing in the straight line, and then recover in a circle; this is done by putting the right foot to the ground when you make the extension, lifting the left foot, carrying it in a circle out of distance, and then recovering the right foot close to it. This is the true method of proceeding and is sure to make a hit.

149

Pl. 59.

This plate illustrates a guard in <u>tierce</u> at an angle, although there is much uncovered above the sword, yet the dagger defends that part; therefore the dagger is held erect and high, with the hand somewhat low and to the outside, so that the adversary can hit only in that part which is exposed above the sword. The intention is to parry with the dagger held thus high and to hit in <u>tierce</u> under the adversary's sword, with the same angle. This stroke is very difficult to parry for it cannot be thrust downwards because of the angle which pushes away the dagger, resists and goes to the body. Therefore the adversary must push your sword to the one side or the other according to which side it is nearer; or rather he must push the sword to that side where it will be out of line in the shortest time, and where it has less force and can offer less resistance. With this guard excellent strokes can be made under the dagger on the outside, and also between the weapons. Sometimes you must lower the point a little and make the angle again hitting with the force of the angle, so that even with both his weapons, the adversary can not push your sword down. You will succeed because his weapons will glance towards the strength of the angle, that is the hilt of your sword. Therefore the adversary has to use much judgment in deciding to which side to push your sword, in order to defend himself more easily; otherwise he must save himself by retiring. That plan of retiring should be adopted, since the thrust at an angle

does not reach very far.   The best rule then is not to parry.

Pl. 60.

This is a guard in _tierce_ with the sword and dagger. It is formed with the points of the weapons in conjunction, so that the adversary cannot hit between them. The sword should be held with one fifth of its length beyond the dagger in order to prevent the adversary from harassing the dagger by disengagements and feints. As the sword is so far advanced, the _forte_ of the sword can be very well used in the defence, with the support of the dagger. There is the further advantage that the point of the sword is nearer the adversary's body, and, if kept free, is ready to hit with a lunge or a pass.   You can wait or attack at pleasure, make feints, disengagements, counter-disengagements, _times_, or _counter-times_; you can lunge or pass, above or below the dagger, in any part according to your opportunity, and with little change of the hand, since it is between _seconde_ and _quarte_.

151

Pl. 61.

This is a guard in _tierce_, with the sword advanced, the dag
ger near the hilt, the right side forward and the left exposed.
With this guard you may advance gradually within distance, in
order to entice your adversary to hit at the part uncovered and
to defeat him at the moment of his advance with a _tierce_ or
quarte; the _quarte_ would be better, because your weapons would
be closer together so that he could not hit below, whilst defending above. When you are approaching within distance if your
adversary changes the position of his dagger, with this guard you
can make a swift thrust, which would hit under the line of the
dagger on the inside and close to the sword; or you might make
an excellent stroke over the sword in the straight line. In this
position the dagger cannot be much harassed by the adversary's
sword, for too much judgment and circumspection would be needed
to bring the sword so far forward. In using this guard you should
know how to preserve your sword from being engaged by the adversary's weapons, when you would have difficulty in freeing it and
could only do so by withdrawing it, which would be bad. Also
you must keep within wide distance, so that your adversary cannot
pass. With this guard it is sometimes advantageous to change to
quarte.

152

Pl. 62.

Here follows another _tierce_ with the dagger advanced and the sword as much withdrawn, in order to keep it free from the adversary's weapons and to give more impetus to the lungs. The sword does not impede the dagger in seeking the adversary's point, and the whole defence is based on the dagger. It is considered that the distance from your dagger to your body is so great, that the adversary's sword cannot penetrate so far, before you have had a chance to parry. This would be true, if he had begun to hit before the point of his sword penetrated your dagger. But against an opponent who knew how to carry his sword forward, so that its point began to penetrate your dagger hand, and knew how to form his decision according to the opportunity, you would be deceived in your judgment, especially if your opponent decided not to advance his sword; to meet that case you must hold your sword advanced and bring the point close to the dagger hand; but promptness is required lest your adversary should engage your sword, and in order that you may take advantage of his movements. This guard is better used in practice, in order to learn the use of the dagger, rather than for anything else. In actual fighting another position and a different style would be required.

153

Pl. 63.

Here follows another <u>tierce</u> formed on the left foot, with the body bent, the sword hand opposite the left thigh and the dagger extended and so high, that the head is almost entirely hidden beneath the line of the left arm. All these positions are to facilitate the defence with the dagger and for the greater security of the body, which is seen to be bent so as to shorten the line of the left side and so that it may be defended with less movement. With this guard the head is low, so that in parrying below there is less exposed above. The hilt of the sword is kept near theknee in order that you may use the <u>forte</u> in defence and be better able to hit. The dagger cannot be much harassed. With this guard you can close distance, or wait and give your adversary various opportunities by movements of the body and weapons. When your opportunity to hit comes, it is better to pass than to lunge.

154

<u>Pl. 64.</u>

Here is another <u>tierce</u> with the right foot carried outwards, the knee bent and the body supported on that knee. The sword hand is advanced towards the adversary a distance equal to the length from the elbow to the hand, and the sword is inclined up-wards at an angle in order to be in conjunction with the dagger, which is held high and in a straight line from the shoulder to it's point. In approaching the adversary the left foot is brought up to the right, and the right foot carried in a circle towards his left side. That part of the side which is seen to be uncovered below the dagger is always kept back, so that if the adversary tried to hit it, you vould parry with the <u>forte</u> of the sword, which is held forward. You should/with the left foot, turning the hand into <u>quarte</u> and always keeping the dag-ger steady with the intention of defending, the upper parts; the body too should be kept at the same height. If you do not wish to pass or cannot pass because you are in motion, you can still parry and hit in the same manner by carrying the right foot into the straight line and without moving the dagger,so that your adversary cannot make a feint below and hit above, or vice versa, and from whichever side he comes you are de-fended. If he disengages over your sword, your dagger, which is steady, can easily parry and defend. If he tries to engage your sword with his dagger, then you can hit under the arm or over the dagger, by advancing in a circle towards that part in order to withdraw out of his line and to hit better; even if

your adversary closed the distance he would effect nothing. With
this guard, if the points of your adversary's weapons are separated, you may make some very good strokes in between as oppotunity offers,if they are close together you can harass his dagger. But with this guard you should never change your hand,
lower your point much, make _appels_ nor stand still, but you
should approach in a circle, keeping the same front as you gradually approach, and form your decision with swiftness. Sometimes
you will parry with the _forte_ of the sword, glide your dagger
along the adversary's blade letting it remain there and hit with
great force. This guard is sufficiently good, though laborious;
but its limitations should be realised.

156

Pl. 65.

This plate is also of a _tierce_ but little practiced, though known by some.   It is formed with the feet together, the body bent, the weapons divided, the dagger and sword both high.   The sword is held high so that it cannot be engaged, the dagger high and withdrawn so that it cannot be harassed, the weapons divided so that the adversary can hit only between them, and the feet together to form a centre with the adversary on the circumference;   at the centre you can change your front by a small movement more than your adversary on the circumference can by two steps.   Moreover with the feet close together you can make a step forward and obtain a long reach, so that with this guard you get within distance before your adversary by an amount equal to the distance between his feet.   Although you are within distance, your sword cannot be reached and you will easily keep it free.   Being within distance, if you wish to pass with the left foot, your dagger will reach the hilt of your adversary's sword, and your body will be turned and go out of line and you will be so far advanced that your adversary cannot free his sword.   If with this guard you advance with the right foot, your sword will reach his hilt before he is within distance, and you can protect yourself on any side by getting out of presence. The power to reach where your adversary cannot reach is a great advantage, coupled with the fact that he cannot engage your sword until he is within

close distance.   Therefore if your adversary remains steady

within presence, he will be in great danger of being hit for

the sword falls downwards with greater speed than that with which

it can be raised upward, as happens with all heavy objects,

When the sword is held high in this manner, before it arrives

in line it penetrates so far that it forces away the adversary's

weapons, and the hit is almost inevitable because of the dis-

tance the sword has penetrated and because very great strength

is needed to resist and check its weight, and also because it

is unexpected, since the adversary, who is not within distance,

is deceived in supposing that you also cannot reach. With this

guard also you can hit in every part,even between the weapons,

with or without a pass after hitting you can recover at once,

but somewhat in a circle.

In meeting this guard you must consider that your adversary's

sword has to fall, and that in approaching you must close its

path, so that it cannot reach the body; you must begin at a

distance, so that your adversary is not within distance even if

he advances his foot. You must hold your sword at the same

height, so that approaching your adversary's sword your point

is near his without your having to move either inside or out-

side. You should advance by natural steps and with the dagger

in conjunction with the sword for greater safety, and you

should move your feet and body towards his right side,in order

to arrive more quickly within his point and reach his body. In

case your adversary on your advance makes any small movement to

change his front, you should then seize that oppotunity and ad-

vancing your feet and body towards his left side hit with a guard of
quarte, holding the dagger high in order not to neglect the defence, although your adversary cannot hit as he has turned both
his feet and changed his front, so that his sword has made a
movement opposite to that needed for a hit. But your greatest
danger is that your adversary, who has his feet together, may
carry back his right foot and leave his dagger for the defence
in such a way, that his body will be turned out of line and his
sword so far withdrawn, that he can return it even if you have
reached his body; but if in that case you left his sword abandoning your first line and in the same time passed behind his
left side, with your head outside his dagger arm; in this way
you would save yourself.

159

Pl. 66.

This is another extension in <u>tierce,</u> in which is shown the position of the dagger, when making a hit, even if there is no occasion to parry, but simply for the safety of the upper part, and in order that the dagger may be ready for all cases. You may also see the manner of bending and lowering the body in order to extend the thrust, preserve a better union of the weapons and offer a smaller target. When you withdraw with the weapons thus united you are better covered. You should always recover the sword close to the dagger and remain at the same height in order to be readier to retire and be better defended; you should maintain the <u>tierce</u> which is advantageous and very safe.

160

Pl. 67.

Now we have a guard in _quarte_ with the sword arm forming
almost a straight line from the elbow to the hilt. The dagger
is not completely extended but in close conjunction with the
sword, the right side advanced in order to keep back that part
exposed above the dagger owing to the dagger arm not being extended. This part is defended if need be, either by the _quarte_
itself or by extending the dagger. Feints above with the idea
of hitting below, or vice versa, do no harm, since the dagger
defends the one part and the sword the other and the sword hits
at the same time. The part outside the sword is somewhat uncovered, but the guard is strong and may easily defend, as the
hand is high. If your adversary desires to hit he will be
forced to pass your _forte,_ where he may be deceived. Therefore
he must attack over the dagger and in the _time_ of your defending that part pass with his sword over the point of your dagger
and hit between the weapons through your _faible_ where the sword
is less strong than elsewhere. This guard is in reality very
convenient for making feints and disengagements with swiftness;
with it you may make excellent strokes between the weapons with
different _times_, but you cannot attack much over the dagger, except by extending the _quarte_ well under the dagger and then
with a disengage turning to _seconde_; in this way your sword
will penetrate and be looked inside his arm in such a manner
that he cannot thrust it away. This guard can be very successful, since the body is well covered, and the sword ready in its

movements and kept free without trouble.

Pl. 68.

This also is a _quarte_ formed with the hands advanced and close together, and the point of the sword so low that the adversary cannot engage nor find it with his dagger. If your adversary also tried to lower himself he would be hit, nor would he succeed by covering your sword with his, since you could easily free yours. The sword in this position defends below while the dagger defends above, wherever the adversary attacks. You can also defend all cuts of _mandiritto_ directed at the leg towards the right thigh, and hit in the same _time_. This is the true method of defending against such an attack. This guard is in fact not useful for the assault, but in defence is very good and safe, for, as the weapons are so close together and extended from the body, your adversary can hit only by passing the _fortes_ of your weapons, which are even stronger by the union of the hands and their distance from the body. But, as we have said, you cannot inflict much damage in the assault, and if your hands can be separated you can be hit without difficulty.

Pl. 69.

162

Here follows another <u>quarte</u> formed with the left foot forward. This is very successful and better than the others formed on that foot, because the left side is covered by the <u>forte</u>
of the sword in such a way, that if your adversary advances to
hit, you have only to hit him, since with this guard you lunge
very far even without passing. The dagger merely defends the
upper part. It is held low in order to close the path between
the dagger and the sword, that you may not be hit there; there
is the chief danger, especially in approaching, more than on the
inside or the outside. Unlike the previous guard with this
guard you may attack or wait, lunge or pass according to your
opportunity. On the other hand it has its disadvantages,
since it is not useful for making <u>appels</u>, and it is dangerous
if the adversary closes; in that case, if you have no time
to hit, you should make a feint in order to hold him, and then
hit. If you observe its conditions this guard will serve excellently in attack and defence.

Pl. 70.

163

Now follows an extension in _quarte_, which shows the manner of holding oneself in parrying a thrust or cat over the dagger. In order that this defence may be strong, in addition to the union of the weapons you should turn the dagger hand that it parries with the edge which is usually below, without dropping the point, so that the blow falls towards the hilt of the dagger; thus the defence will be firmer, and by advancing the right side the lunge will reach further, and the circle made by the body will bring it away from the adversary's point, if you are careful to begin and finish the movement with body feet and weapons in one _time_. You should also bear in mind that when you make a hit and have no occasion to parry, it is not good to raise the hands so far as to uncover yourself below. The weapons should be united and remain so when you retire the body with the point directed towards your adversary's sword in _tierce_ or in the same _quarte_; thus you will recover safely on guard.

Pl. 71.

164

We are now to demonstrate the hits with the sword and dagger, an important subject, the study of which is essential than
in the case of sword alone, because now the sword is held further withdrawn,and the body more exposed. It is not only easier
for the adversary to approach, but he has more opening where he
may hit.   Therefore we have purposely made this discourse somewhat longer than the others that the matter may be better understood,and that the student of this may know how to guard himself
with greater caution against the danger. For there is no doubt
that, when you are taking up your guard, that is the best time
for the adversary to hit or seize some advantage.

Here then is the first hit with the sword and dagger, which
is a quarte and is made against an opponent with a guard of
tierce, who has made neither defence nor attack. He has come on
guard too near to his opponent who has seized the time of his
putting his foot to the ground to take up his position and hit
during that pause.   Thus the adversary has had no chance to do
anything, an error, or rather a stupidity committed by many,
who are wont to say that they were not on guard, forgetting that
when they have the sword in the hand they are supposed to be always on guard. You must observe two things, the first not to
advance so far in taking up your position, that the adversary
can reach you in that time, the second, in taking up your position to be careful not to let the body or the feet or the
weapons drop. You should put your foot to the ground quietly

and lightly with the weapons not far from the position which you
intend to take up. As you approach your adversary you should
adjust your weapons in such a manner, that when you reach your
position the weapons have no other movement to make. Proceeding
in this manner you can take up your position within wide distance,provided that you do not drop the body or foot.   Even
though you were out of distance and though your adversary could
not reach, you might still offer him a _time_ in which he would
seize some advantage, and in the same _time_ approach, when you
would be in danger of a hit while trying to free yourself. Even
if you were not hit at once, yet being at a disadvantage, you
would be hit on the slightest movement. The knowledge of how
to advance against the adversary is certainly the first care,
and of great importance since victory generally depends on the
first advantage.

166

Pl. 72.

This hit in _quarte_ against a guard in _prime_ has arisen in this way: you were in _tierce_ on the outside with your sword inclined upwards,in order to cover yourself from the adversary's sword against a hit in that line. Your dagger is held high, so that it could parry, if he disengaged on the inside, by thrusting his sword down to the left. The adversary,seeing you thus covered has tried to carry his dagger to your sword in order to beat it; at that moment you have lowered your point under the line of his left arm, extended your foot and arm changed into _quarte_ and hit in the part he exposed by his attempt to beat your sword with his dagger. At the same moment you have placed your dagger where the sword was before, and turned the point so low that it has pushed his sword out of line. The point of the dagger has been turned downwards, so that if the adversary had attempted to hit with his guard, his point would have struck the ground. Or it might arise in this way: the adversary seeing his sword subjected,has tried to disengage on the inside in order to free it from your sword, carrying his dagger to the other side with the hope of covering his right side; in the same _time_ you also have placed your dagger on his sword, and thus made the hit shown.

167

Pl. 73.

The next _quarte_ hits against an opponent in _seconde_ who has himself meant to hit over the dagger. Both were in _tierce_; you have made an opening by slightly lowering the dagger arm and approaching it to your sword; the adversary seeing the opening has turned from _tierce_ to _seconde_, disengaging his point, and has thrust over the dagger. You, who have given that opportunity in order to entice him to that side, have parried and carried the right side so far forward that you have not only hit, but also assisted the defence by avoiding with your body; your left side has been carried out of line at the moment of your hitting. Further the change of the hand to _quarte_ has brought your sword away from his dagger. Whilst the adversary was advancing without union between the weapons, your sword has penetrated with the _forte_ before his dagger could find it; and thus you have made the hit. Or it may be that the adversary's sword was on the inside, and you have tried to find it with your dagger, in order to force a disengage; he has disengaged and seized the _time_ to make a thrust in _seconde_ above; you have parried and hit with a _counter-time_ as shown.

168

Pl. 74.

Now follows another hit in _quarte_ against a cut of _mandiritto_ against the leg. You were perhaps in _tierce_ with your sword inclined upwards in order to defend against the adversary's sword in _prime_ or _seconde_; he has made a feint with one of those guards of a thrust towards your face outside the sword; with your point inclined upwards towards his sword it has been convenient for your to defend with the sword, keeping the dagger steady. In the same _time_ the adversary has dropped his dagger under the two swords and made a cut of _mandiritto_ at your leg, keeping his head covered. You,who have gone to the defence with your sword, have let the point drop on the outside of the adversary's dagger, turned your hand to _quarte_, thus bringing it close to the dagger hand, and directed the point at the base of his right side under the hilt of his sword. This has given you a defence for your sword has covered the leg and stopped the adversary's hit. But the sword in parrying should be accompanied by the dagger; for if it had not first reached the adversary's body, it would not have had strength to support the shock, and through the disorder into which it would have been thrown, it could neither have parried nor hit. Therefore you can clearly understand how dangerous it is to try to resist the impetus of a cut, and not to reach the adversary's body before the shock comes. For if your sword is found in the air it is sometimes so disordered in resisting that before it recovers the adversary can repeat his stroke.

169

Pl. 75.

The next plate shows a hit in _quarte_ made close to the adversary's arm between his weapons, but low down against a _tierce_ intended to hit over the dagger. Both combatants were in _tierce_; you have made a feint of hitting in _tierce_ against your adversary's right shoulder; he has parried with the dagger and entered with the right foot in order to hit in _tierce_ in that _time_ under the feint; at the same moment you have changed from _tierce_ to _quarte_, and by simply dropping your point have hit under his dagger hand, which has passed in vain; for when he tried to parry your point, it had already dropped. You have also turned the point of your dagger downwards and parried his sword, whilst your body has been carried out of line in the extension. The arm has remained high, showing that, although you parry below, you should not drop the arm because the _time_ would be long and would cause danger above. It may happen that you have found yourself with your point over the adversary's dagger and in _quarte_, and have disengaged over his dagger with a feint of hitting his right shoulder; he has parried with his dagger and lunged in _tierce_, thus dividing his weapons and leaving an opening between his two hands, so that by simply dropping your point you have hit the part uncovered.

170

Pl. 76.

Now is shown a hit with a guard of _prime_, made in defending yourself from a cut directed at the head and with a parry in the form of a cross, that is to say with the sword and dagger joined together. Your adversary was in a guard of _prime_ or _seconde_; you have tried to engage his sword on the outside with a guard of _tierce_; the adversary has seized that _time_ and tried to hit with a cut of _mandiritto_ at the head, while you were trying to engage his sword. You, being in _tierce_, have brought the hilt of your sword and your dagger together, raised your hands in a cross, and parried with complete protection; in the same _time_ you have thrust the point of your sword towards your adversary's chest over the hilt of his sword, lunged and thus hit with a guard of _prime_; your adversary's sword is excluded between your sword and dagger in such a way that he can free it only with difficulty. This manner of defence is very strong; there is no danger of your weapons being disordered by the shock of the adversary's sword, however great it's impetus; moreover the head is completely defended on both sides at once. Since the lunge is short, you must advance the feet and pass, in order to reach the adversary before he can free his sword. One may say that it is a perfect guard.

171

Pl. 77.

Now follows a low hit in _seconde_ under the adversary's sword also in _seconde_, while he is making a cut at the leg. The plate shows the manner of parrying that cut of _riverso_ at the leg and of attacking in the same _time_. The sword and dagger are shown united in order to add strength to the defence and also to cover the hands and the sword arm against a hit. If the adversary has made a feint of cutting at the leg and tries to hit higher, the conjunction of the hilts of your sword and dagger and the erect position of the dagger will cover all the right side up to the head. If again he has made a feint of cutting at the leg and then cut at the head, your hands will be raised a little, still in conjunction and the body kept at the same height; this you can very well do, because the distance from the leg to the head is so great and the _time_ so long, that you have ample opportunity to defend. The adversary may have attempted this cut of _riverso_ at the leg, when he was in an extended _quarte_ on the inside, and seeing that you were about to hit with another _quarte_ with the point inclined slightly upwards and accompanied by the dagger, he has seized that _time_ to make his cut of _riverso_ at the leg carrying his dagger under your point in order to defend his head; but with your weapons in conjunction you have simply changed the position of your hand, thereby freeing your sword from his dagger, and by dropping your point and body together have hit at the moment of his sword meeting yours. Thus you have defended and attacked in one motion.

172

Pl. 78.

Next follows another _seconde_ against a _seconde_. Your adversary was in _quarte_ with his dagger advanced and his weapons apart; you were in _tierce_ somewhat at an angle and have tried to engage the point of his sword on the inside at the _faible_ , keeping your sword and dagger together. The adversary has seized the _time_ and attempted a cut of _riverso_ at the part uncovered by your movement. With your weapons together you have simply changed the position of your hands, placed the dagger on your sword in order to strengthen the weapons and better resist the shock of his sword, pushed the right foot forward in the same _time_ and made a hit in the chest on the outside of his sword, parrying at the same moment. We have introduced this plate in order to show how to parry and hit against a cut of _riverso_ at the head, and how to resist a blow of the strongest arm.

173

Pl. 79.

This also is a hit in _seconde_ against a _tierce_. You were in _tierce_ and have changed to _quarte_ with the dagger in union whilst engaging the adversary's sword in order to exclude it; and form a counter-position; he has seized that _time_ to disengage and hit in _tierce_ on the outside of your sword. At the same moment you have changed from _quarte_ to _seconde_, keeping the dagger for the defence against any possible thrust below. Or it may be that you had formed your counterposition, and the adversary has attempted a cut of _fendente riverso_ at the head, which was exposed; in the same moment you have changed to _seconde_, advanced, parried with your sword and hit in the same _time_. Although this defence is weaker than a parry with the weapons in union, still it is good, because the lunge reaches further; but it is certainly not so safe.

174

Pl. 80.

Now follows a defence in _quarte_ accompanied by the dagger against a cut of _mandiritto_ at the head. The _quarte_ is directed between the adversary's weapons. He was in a guard of _seconde_ and you have tried to engage in _tierce_ on the outside and with your dagger in union in order to exclude his sword. The adversary has seized the _time_ and made his cut at the head, thinking he could hit the part uncovered. With your weapons in conjunction you have simply changed your hands from _tierce_ to _quarte_ and parried with the edge of the dagger which is generally below; for in this way, as we have already said, the defence is stronger, the adversary's sword kept further away, and the hands more united, you have also brought the _forte_ of your sword into that line, which not only defends the head and makes the parry safer, but also covers the part below the dagger, so that, if your adversary had made a feint of cutting at the head and changed to _seconde_ in order to hit the left side, while you were raising your dagger to parry, the _forte_ of your sword in this position would still have defended your left side. In addition to these two good results you would also have hit in the same _time_ for your adversary could not have parried without abandoning his plan and changing to _seconde_; this would have changed the front of his body, so that he could have defended himself, but could not have hit, since you would have covered all that line.

175

Pl. 81.

This plate illustrates a hit in _seconde_ over the dagger.
Both were in _tierce_ ; you have turned your sword downwards and
the adversary has changed his hand to a low _quarte_ in order to
hit between the weapons, and remain close to your _faible_, keeping his dagger low In order to be more covered below, although
he would have been better defended with the point of his dagger
You have recognized his error, seen the opening above, changed
your hand from _tierce_ to _seconde_, disengaging your sword over
his dagger and hit in the part uncovered. By placing your dagger on his sword you have easily defended, aided by the alter-
ed position of your body which has been carried out of time.
Or it may be that you have directed your sword at the adversary's
dagger hand,and he has tried to engage it and hit in _quarte_ with
the idea of closing the path between the weapons; you have then
disengaged and made the hit. The adversary has been unable to
parry, though he raised the point of his dagger, because your
sword was shut in between his arm and dagger.

176

Pl. 82.

Now follows a hit in _tierce_ against an opponent who has attempted or hit in _quarte_. You were in _tierce_ with the right
foot in front, the feet close together, and the weapons in union;
you have deliberately moved aside the point of your sword and
left an opening between your weapons, your adversary seeing this
has thrust in _quarte_ close to your sword in order to cover himself while hitting, you have made a _counter-time_, and by lowering your sword hand and body and carrying your dagger, which
had not so far been moved, to his sword and engaging it, you
have hit in _tierce_. This result might have arisen in another
way; you have made a feint of hitting in the straight line
over your adversary's dagger; he has tried to parry with the
dagger and hit the right side of your chest in _quarte_. Then
you have dropped and parried; by putting the point of your
sword under his dagger you have made the hit shown.

177

Pl. 83.

This is another quarte, which has parried a seconde with
the hilt of the sword. Your adversary was in tierce with the
point high and the hand low; on your advance he has disengaged
in order to hit in seconde the part exposed under the dagger,
and has tried to place his dagger against your sword in order
to defend himself. In that time you have changed your hand
from tierce to quarte, and keeping it close to the dagger hand
have carried it past the point of his dagger and reached the
body, hitting his right side. By raising your hilt during the
stroke you have defended the part, which was exposed, and where
your adversary intended to hit; your dagger has been left to
guard the lower parts in case of a hit there. Or it may be
that your adversary has made a feint of hitting in seconde over
the dagger with the intention of hitting below; you have parried and thrust, changing your hand to quarte and accompanying
it with the dagger, which has defended above, whilst the sword
has defended below, so that your adversary has been deceived;
he has been able neither to hit nor parry; since your sword
hand was raised, your sword has escaped his dagger, which has
fallen after missing your sword owing to the angle it made, which
has brought it away from the dagger. This hit also shows the
importance of the union of the weapons.

178

Pl. 84.

This plate presents a hit in seconde against a quarte.
Both combatants were in tierce; you have made an opening between your weapons by carrying the point of your sword away,
keeping the sword hand steady and the dagger also. In that time
your adversary has thrust between your weapons, forgetting that
he was within wide distance, and that you had not moved your
feet, and that therefore he could not reach before you had finished your movement. Thus he has been hit. You have offered
the opening and, seeing him coming, have turned from tierce to
seconde, changing the front of your body; placing your dagger
against his sword you have parried, advanced the right foot and
thus hit in the part uncovered by his lungs. Or it might arise
in this way; your adversary was in tierce and you in seconde
you have made a feint over your adversary's dagger; he has
been deceived by the feint, tried to parry and hit in quarte at
your chest while your were approaching. You have parried with
your dagger, which was steady, disengaged the point of your sword
underneath his dagger arm on the outside, and by this path have
made a hit in the left side. Whether the hit has arisen in the
one way or the other, it is certainly due to the separation of
the adversary's weapons; if he had moved with his weapons in
union, although he might not have hit, he would still have been
defended on both sides.

179

Pl. 85.

Now follows another hit in _seconde_ between the weapons
against a _quarte_. Perhaps you were in _seconde_ and your adversary in _tierce_, his sword advanced and dagger close to the hilt
of his sword, with his left side kept back in order to offer
you an opening to hit, with the intention of parrying and hitting in the same _time_, you have feigned to accept the opening
and hit where he desired; thus he has been deceived by your
trick, raised his dagger,and advanced his right side still
further in order to hit,and to defend at the same time below
by changing his hand to _quarte_. You have disengaged over the
point of your adversary's dagger and hit between the weapons
in the part exposed by his attempt to parry and hit. You have
left your dagger in its original position, which has given you
a defence, and excluded his sword in _quarte_ on the outside. If
your dagger had met his sword further forward,it could not have
thrust it away, since the _quarte_ is very strong in that part,
with the result that both would have been hit. If you had
tried to attack under the dagger,your adversary would easily
have parried by merely making a somewhat larger angle with
his sword hand, for his body was already sufficiently turned.

180

Pl. 86.

Now follows another hit in _seconde_ over the dagger against a _tierce_ meant to hit below the sword. It has arisen from your making a feint of hitting in _quarte_ between the weapons; your adversary has tried to parry with the dagger and hit in _tierce_ below; you have changed from _quarte_ to _seconde_ raising your sword past the point of his dagger, and have hit in the _time_ of his attempted parry and thrust; moreover you have defended below by placing your dagger on his sword and pushing it outside your right side, at the same time bending your body. Or it may be that you were in _tierce_ above the adversary's dagger and have made a feint of hitting between his weapons, which were also in _tierce_; he has tried to parry and hit together; by bending the body and advancing the right foot you have met his sword with a _counter-time_, since you have returned your sword above to the same line as before, changing from _tierce_ to _seconde_, which has facilitated the defence of the dagger.

181

Pl. 87.

This is also a _seconde_ hitting over the dagger against an opponent in _quarte_, who has tried to parry with his dagger and turn his body with the left foot. The adversary was in a low _tierce_ somewhat advanced with his dagger close to the hilt of his sword; you have entered over his sword with your hand towards _quarte_ and the dagger also in the same line, so that your adversary, with his sword advanced in the manner described, could not thrust on the outside or below, but was forced to approach on the inside only. You have closed the distance and your adversary has tried to carry himself outwards in order to save the part uncovered, when he has in the end been hit; he has expected that you would disengage below his dagger hand and hit above and therefore he has turned in order to hit in _quarte_ His move might have succeeded, if you had not deceived him. After getting within distance, you have waited for him to move, and seizingthat _time_ have disengaged from the point of his dagger, changing your hand to _seconde_, and thus have hit in the weak part. Although the adversary has turned his body and tried to parry with his dagger,he has done no good, because he has been too weak; nor has he hit you, since you have carried your dagger to his sword and turned the body in hitting; by carrying your right foot somewhat towards the inside you have easily defended yourself.

182

Pl. 88.

This _quarte_ with a turn on the left foot has parried a stroke over the dagger in _seconde_ and at the same time hit the adversary between his weapons. The adversary had his dagger somewhat advanced and has tried to parry, but had been antici- pated by your sword which has reached his body first; he has encountered the _forte_ of your sword with the point of his dag- ger and therefore has been unable to defend himself. You had moved to engage the point of his sword with the point of your dagger, and he,who was in _tierce_,had disengaged in order to hit in _seconde_ the part uncovered above; but you had made little movement and have defended before the _forte_ of his sword penetrated. You have succeeded because your dagger induced his sword to move; if on the other hand his sword had caused your dagger to move, his thrust would have arrived while your dag- ger was falling, so that you could not have parried.   In the actual case his sword has moved, your dagger has been raised and parried, as shown, with the help of the turn of the body.

183

Pl. 89.

This plate shows a <u>tierce</u> which has hit under the dagger against an opponent in <u>tierce</u> with his left foot forward. You were in <u>tierce</u> in a straight line with the point of your sword opposite your adversary's dagger hand. He has tried to get within distance with his left foot, and you, who had your point near his left hand,with a very short movement have thrust under the line of his arm and so close to it, that in hitting your sword has passed out of his view; he has been able to see only that part from his dagger hand to the hilt of your sword, you have held your dagger steady with the point towards his sword, and your dagger has approached just so much as his foot has advanced, and thus has been ready to parry in case of need; but the adversary has made no movement of defence owing to his being surprised whilst moving his foot. The same hit might have arisen in this manner: you had the point of your sword against the adversary's dagger hand and have moved the point towards the inside; he has tried to approach with the left foot in order to engage your point and to close that line somewhat; you have seized that <u>time</u> and made the hit shown. Also if you had withdrawn so far that your adversary could not reach you except by passing in this manner you could have retreated without danger.

184

Pl. 90.

The next is a hit in _quarte_ against a guard of _tierce_. It has arisen in this way: you were in _tierce_ with your left foot forward and your weapons together; the adversary has attacked with the point of his sword outside your dagger; in advancing he has penetrated your dagger with the point of his sword intending to subject it and separate your weapons, in order to hit between them or under them at your side. You have taken that _time_, partly turned your dagger in order to parry with the edge, turned your sword hand to _quarte_ at the same time and thrust so close to his sword that he has not been able to put it where he intended, nor to free it. Or it may be that your adversary has been pressing you on the inside when you were in _tierce_ on the left foot and with your weapons together; you have turned the points of your weapons against his sword to exclude it on the outside; then he has disengaged for fear of his sword being engaged. You have moved simply to make him disengage; you have taken the _time_, lunged with your weapons in union and penetrated as shown a distance equal to the length of your step. Here you may realise the force and great impetus of a stroke made with the left foot forward, and followed by the right foot.

Pl. 91.

185

Now follows a hit in <u>prime</u> over the dagger of an opponent in <u>tierce</u>. You also were in <u>tierce</u> with your point so low as to form a straight line, your feet close together and your dagger in the line of the adversary's sword. He has advanced the point of his dagger low down in order to find your sword; taking the <u>time</u> you have disengaged over the point of his dagger, turning your hand from <u>tierce</u> to <u>prime,</u> and by carrying yourself forward at the same time have hit in the chest. If you had disengaged your sword under his dagger hand, you would not have hit, because his dagger arm being so high would have covered all that part of his body. Your dagger being in the line of his sword has also advanced, as you carried your body forward, and arrived close to the blade of his sword ready to parry, if he had tried to hit. You have not beaten his sword in order to avoid the risk of his disengaging and in order to keep the dagger steady. Or it may be that you were in <u>tierce</u> with your point close to the adversary's dagger and have gradually moved it to the inside and somewhat low in order to induce him to follow it and engage. This has happened, so that with this trick you have drawn his dagger out of the line and caused him to expose himself over the dagger; thus you have suddenly made the hit shown. Often it is better to proceed in this manner, when your adversary tries to engage your sword, rather than to disengage. For by disengaging generally you do what your adversary desires; by not disengaging and adopting this method of

saving your sword, you cause your adversary, if he follows it, to uncover himself; he is disconcerted and can no longer defend himself. If he does not follow it, your sword is free and the danger past.

Pl. 92.

Here is a hit in seconde over the dagger against a tierce formed with the left foot forward. You were in tierce with the feet close together and the point of your sword in line with your adversary's dagger hand. He has pushed forward with his left foot in order to engage your sword on the inside with his dagger. In that moment you have turned your hand from tierce to seconde, advanced the right foot and made the hit over his dagger. You have kept your dagger extended and close to the point of his sword, so that you could have parried with little trouble, if he had tried to hit. Or it may be that you had your point in line with your adversary's dagger hand, and had dropped the point somewhat; he has followed with his dagger and advanced his feet in order to engage it; in that instant you have dis-engaged, turned to seconde and hit in the time of his following your sword; therefore he has been unable to parry.

187

Pl. 93.

The next hit in seconde against a seconde may have arisen in several ways.   In the first place both were in tierce on the outside; the adversary has tried to force your sword by turning his hand from tierce to seconde in order to strike the sword on top, and to parry below with his dagger. Aware of the force of his sword you have eluded it, and by turning your hand to seconde have hit on the inside over the point of his dagger. In the same time by a turn you have carried your body away,and by bending the right side,which was in danger, and bringing the left side forward, have brought your body well away from the adversary's sword. By carrying your dagger to his sword at the same moment you have parried, as shown.   Or it may have happened that you were on the outside and have made a feint of hitting in tierce over the adversary's sword; he has turned his hand to seconde in order to parry and hit in the same time,protecting himself below with his dagger. In that time you also have disengaged, changed your line and hit as explained, parrying with the dagger. By bending your body you have let his sword pass wide, for the hit in seconde falls naturally of itself when it meets no resistance. Or again you may have been on the inside and found an opening between your adversary's weapons; you have made a feint in quarte in that line close to his sword, and your adversary, raising his hand from tierce to seconde and putting his dagger on your sword in order to defend that line, where it had entered, has pushed on in order to hit in the same

time.   Your hit in <u>quarte</u> was a feint; you have carried back
the right side which was in front and raised your hand from
<u>quarte</u> to <u>seconde</u>;   thus your body has been brought out of line
and your dagger has easily defended.   Your sword arm has yielded and eluded his dagger;   he has failed to find your sword with
his dagger, since it has been raised and has hit without impediment over the point of his dagger.

189

<u>Pl. 94.</u>

Next follows a <u>quarte</u> hitting on the inside through the <u>faible</u> of the adversary's dagger, who is in <u>tierce</u> on the left foot. You also were in <u>tierce</u> in the straight line with the point of your sword opposite his dagger. Whilst closing distance you have kept your sword hand in the straight line, until your adversary was within distance. Your feet were close together and your point in line with his dagger hand. When he had advanced within wide distance with his left foot, seeing that his point was somewhat divided from his dagger, perhaps in order to prevent a possible attack above, you have changed your hand from <u>tierce</u> to <u>quarte</u> and thrust through the <u>faible</u> of his dagger. He has failed to parry, since this guard of <u>quarte</u> is particularly strong in that line; further in lunging your sword has come so near the edge and <u>faible</u> of his dagger, that you have not only made his parry fail, but pushed it away by main force and almost knocked it out of his hand. The angle which the hand forms in this guard if <u>quarte</u> has carried the point to the adversary's left shoulder. Or it may be that you had your point against his dagger hand, and on your moving it towards the outside, he has carried his dagger in that direction, raising it in order to cover himself; on that small movement of his dagger you have disengaged over the point and made the hit.

190

Pl. 95.

This plate represents a hit in <u>tierce</u> under the sword. You were in <u>tierce</u> with the left foot forward, and your adversary in <u>quarte</u> with his point inside your dagger, and within such close distance that his point has penetrated your dagger. In the same <u>time</u> you have placed your dagger on his sword and gliding along his blade have advanced the right foot; by dropping your body you have hit between his weapons, but so low, that he has found no defence, and has been caught unawares by the movement of your foot. Or it might be that you, when on guard, had moved the point of your sword outwards, your sword being inclined upwards. Your adversary has tried to exclude your sword with a <u>quarte</u>, advancing his foot. Then you, being steady on your guard, have taken that <u>time</u>, dropped your point, so that he has failed to find it, and have passed with your body bent as low as before, reaching your adversary's body by advancing the other foot. The same hit may have occurred in another way: you were in <u>seconde</u> and your adversary in <u>tierce</u>; you have made a feint in <u>seconde</u> over the dagger, he has raised his dagger to parry and changed his hand to quarte in order to hit also; at the beginning of his movement you have dropped your point, turning from <u>seconde</u> to <u>tierce</u> and engaging his sword at the same time; you have passed and made the hit at the moment of his trying to defend. In addition to the other reasons this hit has followed because the adversary has moved his foot and sword, when you have moved

your point only, so that your dagger has engaged his sword and
he has been unable to save it. From this you may learn, that
on moving your foot you should not advance the point of your
sword so far that the adversary's dagger can get control of it.

Pl. 96.

The last plate with the sword and dagger represents a hit
and pass in _seconde_ against an opponent in _tierce_ who has tried
to parry with his dagger and failed,because the _forte_ of your
sword has penetrated too far; in that part your sword is
stronger than in any other. This hit may have arisen in several ways; in the first place you were in _seconde_, on the outside and your adversary in _tierce_; you have moved to engage
his sword with your sword in _seconde_ and accompanied by the dagger; he has tried to disengage to prevent his sword being subjected, with the intention of advancing and hitting; with your
weapons in union you have changed the line of your sword and
dagger and the front of your body advanced the left foot and defended yourself; by bending the body as far as possible you
have hit between his weapons. Or it may have arisen in this
manner: you were on the inside and have engaged your adversary's sword which was in _tierce_; he has left an opening between his weapons and on his making some movement with his
foot you have taken that _time_ and hit. Or it may be that after

engaging your adversary's sword you have made a feint of passing
and hitting over the point of his dagger; you have forced him
to raise his dagger or move it, and taking that _time_ you have
returned between his weapons and hit in _quarte_, passing as
shown. Or it might well be that your adversary had engaged your
sword; you made a feint of disengaging over the sword and have
returned to the inside in the time of his attempted parry; putting your dagger under his sword as he raised it, you have
thrust and hit. The stroke might have happened in yet another
way: you were in _tierce_ with your sword free on the inside and
by making a feint of hitting over your adversary's dagger have
separated his weapons, and in that instant by passing with your
sword over the point of his dagger have hit between his weapons,
excluding his sword on the outside with your sword and dagger
in conjunction; by the extent of your advance his sword from
being straight has been forced into an angle .

193

PART 3.

GENERAL DISCOURSE ON THE SWORD & CLOAK.

Fencing with the sword and cloak is a noble art, and in general use in every province where it is not prohibited. But like the dagger in many states and cities its use is forbidden by the ruling princes, so that it falls into disuse, and the labour and time spent in acquiring an understanding of the art are rendered vain. However, we have considered it fitting to treat of it in order to explain its nature, and how and when it should be employed.

We say thus that the cloak is both a defensive and offensive weapon; by offensive we mean capable of inflicting damage on the adversary. By being cast in various ways it can impede his view and his hand, a disadvantage which the user of the cloak may himself suffer by throwing it over his head and impeding his own view; though we think that this should happen only to one entirely inexperienced in its use. To one who understands it well, it is a very advantageous weapon. Its use requires an understanding of the sword alone, since in many cases the sword defends with the aid of the cloak, especially against cuts at the head. In parrying such cuts you should never interpose your arm, owing to the danger of the cloak being cut and the arm hit. Even if you wrap the cloak round your arm, besides leaving the lower parts exposed with grave danger, it would not support such a stroke without injury to the arm. Against

such cuts it is beyond comparison better supposing that the
cloak is held correctly, to bring the left foot forward and, extending the cloak arm, to parry near the adversary's hand; in
this way the danger would be avoided. In case you were not
near enough for such a move, you could let the cut pass without
resistance, and then advance the sword and cloak together.
Otherwise you should parry with the sword supported by the cloak
or with both together, hitting in the same _time_.

The lower parts can be very well defended by the cloak in
spite of its weakness; for it yields to the shock and in addition has its length and width. These three conditions give it
its power to resist and parry. With its flexibility and without
its width it would not defend, however long, so that its strength
consists in its width; but the part which yields must be aided
by the movement of the feet and body to render the defence secure both against cuts and thrusts. The cloak also will defend
all thrusts under the arm on either side by its edge hanging
from the hand towards the ground. The arm must be held extended
with the hand towards the adversary in order to parry far from
the body and in order that his point may not reach your body
on the yielding of the cloak before it is driven out of the
line. Therefore you must not hold the width of the cloak facing the adversary, lest he should hit in the middle, for it is
more difficult to parry with the cloak in this position than
when it is held on edge; in the later case it easily carried

the thrust outwards on either side. Besides holding the arm extended you must hold it so high that your hand is on a level with the head whilst you look towards the adversary along the line of that hand. The cloak must hang only so far, that on lowering the arm, whether from fatigue or any other reason, it does not reach the feet with the danger of causing you to fall . You should cover your arm with the cloak up to the elbow, and the point of your sword should be in conjunction with the cloak hand, for better protection and for the strengthening of the sword, and in order to defend the cloak hand better. When your arm is weary you should bring it towards the hilt of your sword and close the two hands so that the adversary cannot come between them; you should face in such a way that the edge of the cloak always looks towards the adversary.

When a thrust is made in the upper lines, you should parry by raising the hand only, leaving the elbow steady in such a way that from the hand to the elbow a perpendicular line is formed.   If you observe this rule you will force your adversary not to lift his sword but to attack on the outside, where you may protect yourself more easily and with less movement. If in the time of your parrying the adversary tries to hit in the centre, you will defend yourself more readily, since there will be less exposed than if you had moved the whole arm. In this way the elbow forms the centre, in the other the shoulder, hence in the latter case the movement of the hand is greater,

the part uncovered larger and the defence of the face more dangerous if the adversary's sword is raised, and this because the
path taken in driving his sword out of line is longer. If your
adversary attacks on the outside of the sword towards the face,
still you should not move your elbow but keep it steady,raising
the hand only in such a manner as to defend that part to the
top of the head. This will give you two advantages; firstly
a better defence for you will not only parry with the cloak
hand, but also make use of it and the arm down to the elbow,and
so cover the whole of your right side; in the second place you
do not obstruct your view, whereas if you raise the whole arm
you obstruct your view and cannot see the adversary; if you
observe the rule you will always see the adversary's sword hand.

If the cut at the head is in _mandiritto_ you should parry
with the sword by turning the hand to _quarte_ and thrusting the
point towards the adversary's chest or face, holding the cloak
hand close to the hilt. If you do not hit with the parry you
should at once turn your hand to _seconde_ or _tierce_, carrying
the cloak hand to your adversary's sword on the one side or the
other, according as it is more on the inside or the outside,
then bring the rear foot forward and hit in the part uncovered .
With this parry you can also make a feint with the point and,
after parrying,make a cut of _riverso_ at the leg, leaving the
cloak arm against the adversary's sword and bringing the rear
foot forward; if ou act swiftly you could also make a thrust

at the chest. But if you do not wish or cannot make this defence, you should parry by holding the cloak hand under your
sword, and after parrying at once disengage your sword in
seconde,and leaving the cloak below the adversary's sword hit
in the chest; or after parrying cut in _mandiritto_ at the leg;
or still defending yourself with the cloak above, cut in _riverso_.
If your adversary cuts in _riverso_ you could parry with the
guard of _seconde_, thrusting towards his chest at the same moment, and leaving the cloak hand close to his sword for the defence of the lower parts; if you have not hit with the parry,
you should leave the cloak against his sword and hit below in
tierce, or you must immediately after parrying cut in _mandiritto_
or _riverso_, still letting the cloak defend above. If your adversary should cut in _mandiritto_ at the leg, you could parry
with the cloak and hit above with the hand in _quarte_, so that
if his cut at the leg has been an artful feint and he had then
cut at the head, the sword would have defended. If he should
cut in _riverso_ you should parry still with the cloak,and with
the sword in a guard of _seconde_ for the defence of your head
advance to hit in the same _time_; if on your right foot you
must carry it towards the left in order to cover yourself
more by the cloak, and also lest the adversary should reach you
by the yielding of the cloak; you should turn the point of your
right foot outwards, in order that the cloak arm may approach
nearer the line of the adversary's stroke. But if you pass with
the left foot, you avoid this danger, because the cloak would

cover more, so that you have only to make the hit directly. But if your adversary has made a feint of _riverso_ at the leg and changed to a _mandiritto_ at the head, you should then turn your hand to _quarte_ holding the cloak close to the sword and hitting at the same time. You could also parry with the cloak alone, if you had advanced, because you would have brought it so near to his hilt that there would be no damage even to the cloak; you should hit in _tierce_ below in the same _time_. Or if you were in _seconde_ when your adversary cut at the head, you could cut in _mandiritto_ at the leg, making the same defence, or disengage from seconde and cut in _riverso_ at the adversary's right leg.

All thrusts whether at the head or body, can be parried with the cloak; but thrusts between the hands must be parried with a change to _seconde_, otherwise you would be in danger of being hit; for besides the yielding of the edge of the cloak on reaching the adversary's sword, there is the danger that your right shoulder in its advanced position may be reached by his point before the point is driven out of the line. Your arm too is in danger of obstructing your defence and of being hit, as you do not wish to withdraw the arm while parrying. Therefore by changing from _tierce_ to _seconde_ you make the defence easier, since not only does your arm give place to the cloak, but the front of the body is changed in such a way as to escape the danger, and you hit at the same moment. When his point comes outside your cloak or sword, there is not the same difficulty as

when it comes between the two. Disengagements should be made
above, for the hanging cloak obstructs their being made below.
But on the side of the sword they can be made below, when making a hit outside the sword. But when the adversary's sword
and cloak are separated, it would be well to pass the point of
his sword if it is not inclined upwards at an angle, for then
you could reach the body without bringing your _faible_ near his
forte. If your sword is on the outside of his, you can disengage over the point, and hit in the middle, and sometimes over
the cloak hand. Again, if your sword is over his cloak, disengage in the middle, and if his weapons are united and you cannot hit there, hit over the sword, or make a feint of a hit
there, and, when he moves to parry with the sword,hit in _tierce_
below, parry with the cloak and recover to the position over the
cloak hand with a change to _seconde_; in this manner you will hit
in the same time and defend yourself with the cloak against a
possible stroke below.

These rules apply when on guard on the right foot. When
on guard on the left foot you should proceed in a different manner. The position on the left foot is better with these weapons
than with the sword and dagger, for the parts which are exposed
and generally attacked with those weapons are in this case better protected, since the cloak hand can be held so high as to
defend all the part exposed above the arm, while the hanging
edge of the cloak defends below; thus both parts are covered
at the same time, whereas with the dagger in covering one part

you expose the other, besides the grave danger to the left knee, which is advanced and far from the defence. With the cloak the knee also is defended. Farther the sword in _tierce_ cannot only be brought in conjunction with the cloak, but can rest on the cloak hand and be strengthened in such a way that it cannot be thrust aside. With this union of weapons you can make all the defences and attacks; if you have formed your guard well, your adversary will be able to hit only over the sword, where there is little exposed; your sword, strengthened by the conjunction of the left hand, will parry with ease and without further change. This is impossible with the sword and dagger, since there are more exposed parts, and if you rest the sword on your dagger, you obstruct the dagger and destroy its function. On the other hand such a position strengthens the cloak and gives a better defence. With this guard you have merely to push on within distance of the adversary, where you can hit without separating your weapons. You should observe this union whether defending against cuts or thrusts.

So far we have treated of the defence and attack with the cloak with respect to the difference from the methods with the sword and dagger. Now we must mention that the cloak can be used by throwing it in various ways, such as by wrapping it round your adversary's head, and letting it go entirely, or holding it by the lower edge in order to bring it back to your arm, if the plan has failed. It may be thrown over the adver-

sary's sword, but it must reach the hilt of his sword in order
to check and impede his hitting or doing anything else. You may
throw it round your arm in order to deceive the adversary and
then cast it in his face. You may rest the point of your sword
behind it and then carry it to his face. In all these cases you
should hit before the adversary frees himself. These are the
tricks of the cloak which are not expected by the adversary, and
the unexpected attack is more effective. Here we end the general
discourse on this weapon.

202

Pl. 97.

Here we shall explain the principles of the accompanying plate, the first in order with the sword and cloak and formed with a guard of seconde. We have thought good to neglect entirely the guard of prime as unnecessary, in order not to make the book unconscionably long. When this guard in seconde is well formed it is safer than that with the sword and dagger, because the lower parts are defended by the cloak hanging from the arm. The arm is held high and to the outside, so that the adversary cannot hit above, nor can he hit between because the sword is held close to the edge of the cloak. As there is nothing exposed over the sword, he can hit only the small part exposed below near the cloak, which the cloak can easily defend, especially if aided by the right foot. In advancing you should carry the right foot towards the adversary's right side with the point of the foot towards the left on reaching the ground; in this way the right side of your body will be brought out of line, which will help against the yielding of the cloak on reaching the adversary's sword; thus you will be safe and hit at the same time. In finishing your lunge you should raise the left foot and carry it back in a circle towards the left, bringing the right foot back to it immediately and recovering to the same guard of seconde; you should form this position as often as there is an opportunity to hit.

203

Pl. 98.

The second plate with the sword and cloak represents a guard in _tierce_, with the sword inclined upwards at an angle for two reasons; in the first place to close the path between the weapons, in the second to cover and defend the cloak hand, in case the adversary should try to hit it; therefore the sword is held advanced; also if the adversary comes on the outside of the sword, the _forte_ being well advanced can easily defend. Instead of using the sword you could parry with the cloak alone, change the hand to _seconde_ and hit in the chest in the same time recovering on guard in _seconde_ in the way explained in the last plate. When steady on your guard; if you wish to change from _seconde_ to _tierce,_ you must carry the left foot back somewhat, so that the adversary cannot reach you during the change; afterwards you should begin to close the distance with this _tierce_; if it is well formed the adversary will be able to hit only over the sword.

204

<u>Pl. 99.</u>

The next is also a guard in <u>tierce</u> formed because of the
fatigue of the arm, which cannot be held long extended owing
to the weight of the cloak which drags the arm down. When the
arm is weary you must carry the left foot backwards in order to
withdraw the upper parts most exposed, leaving the sword advanced to keep the adversary at a distance; the cloak hand
should be united to the sword hand to prevent a hit between;
thus the adversary will be able to hit only over the cloak arm,
and against this stroke you must raise the cloak hand only,
without moving the elbow from its position, accompanying it
with the sword hand changed to <u>quarte</u>, in order to close the
path between; thus you defend the upper parts and hit at the
same time, immediately recovering to <u>seconde</u> with the cloak extended.

205

<u>Pl. 100.</u>

This also represents a <u>tierce</u>, but on the left foot. This guard is better with these weapons than with the sword and dagger or any other weapons, because the side and the leg, which are in the greatest danger, are covered by the cloak, which hangs in such a way that neither cut nor thrust can attack. Further the sword hand is held so high that the adversary cannot hit above, while the path between is closed by the sword which rests on and is strengthened by the cloak; the sword is directed towards the adversary in line with your cloak hand in such a way that he has nowhere to hit except in that line,which is defended by the <u>faible</u> of your sword; though we call it the <u>faible</u> it is none the less stronger than the adversary's <u>forte</u> as it is strengthened by the cloak hand; and with this support your sword hits at the same time; with the same <u>tierce</u> you can reach his body and after hitting recover to the same guard. In short this guard is the best of all with the sword and cloak.

206

Pl. 101.

The next plate represents a defence with the sword and cloak with a guard of _seconde_, which is far better than with the sword and dagger, because the lower parts are easily defended by the hanging cloak; if the adversary should make a feint of cutting at the head in order to thrust below, he would do no good, because you would turn your sword to _seconde_ and advance to hit, carrying the edge of the cloak somewhat towards the right so as to be covered; if his cut had been really meant and had been a simple cut, after parrying you would still have made a hit in _seconde_ inside his sword, leaving your cloak on his sword and passing on with the left foot, with the point of the foot outwards, in order that both sides of the body may advance and the sword reach further. This kind of hit will be treated of in its place, that all may better understand it.

207

Pl. 102.

Next we show a hit in _quarte_ against a _seconde_, intended to hit over the cloak arm. You were in _tierce_ and have dropped the cloak arm towards the hilt of your sword; your adversary seeing this has advanced his right foot to hit the part uncovered, changing his hand from _tierce_ to _seconde_ and carrying his cloak to your sword in order to parry; in that moment you have raised your cloak hand and carried his sword far out, changed from _tierce_ to _quarte_, and without allowing your sword to be caught by his cloak and have hit in the right side. This high defence is very advantageous with these weapons, since the cloak covers the whole of the part which is exposed with the sword and dagger; the hand also by the change to _quarte_ covers the inside, so that the adversary could effect nothing there. Since your point hits the base of his arm you may easily realise that he cannot recover into line.

208

Pl. 103.

Next follows a hit in _tierce_ against an attempted hit in _quarte_. You were in _tierce_ on the inside and your adversary has tried to exclude your sword; you have disengaged in low _tierce_; taking that _time_ in order to hit between in _quarte_ he has lowered his cloak to defend the lower parts; in that moment you too have dropped the elbow of your cloak arm, raising the hand so as to cover the whole face and thus entirely closing the path between your arms and have directed your sword on the outside of the cloak arm in the line left by the dropping of the elbow, for otherwise your view would have been obstructed by the cloak; at the moment of doing this you have thrust in _tierce_ at an angle, so that your sword has passed without opposition from his cloak. Or it may be that you had tried to engage the adversary's sword on the outside. Seeing your weapons divided he has tried to hit in the opening by disengaging in _quarte_ past your _faible_. You have advanced your left side, which was behind, and resting your cloak on his sword have pushed it out of line and driven his points so far upwards that it has not encountered the cloak; in this way you have made a hit over his cloak, and all the better if he has perhaps lowered the cloak in order to defend the lower parts, and thus has been unable to parry the stroke.

209

Pl. 104.

In the next plate we see a hit in tierce against an attempt- ed hit in <u>tierce</u>. You were in low <u>tierce</u> and have carried the point of your sword away with the intention of leaving an open- ing between your weapons; the adversary has been enticed by the opening to hit in between, so that he has divided his weapons; his cloak hand has been left behind by his advancing the right side of his body. Being in a low position you have extended your arm and right foot and hit; your lunge has caused his sword to fall low. The same hit might be made in this way: you were in <u>quarte</u> with your arm withdrawn; you have allowed your point to be engaged by your adversary's sword in order to give him an opportunity to hit between the weapons with a disengage on the inside, he has attempted this hit and you, who had moved with this intention, have made a half turn from <u>quarte</u> to <u>tierce</u> which has freed your point and given you the chance to parry and hit.

210

Pl. 105.

The next plate shows a cut of _riverso_ at the leg with the left foot forward. The adversary had made a cut of _mandiritto_ in the _time_ of your moving to engage his sword, which was in _seconde_ on the outside. You, being in _tierce_, have perhaps parried and immediately let your sword fall on his leg, keeping the cloak at the defence. The stroke might be made in this way: the adversary has made a cut; you have parried with a _quarte_ carrying the point towards his face; your point having failed to reach, without an interval you have changed to _riverso_, passing with your sword between your own cloak arm and his sword; leaving your cloak for the defence, you have advanced the left foot and made the hit. If you had wished, you could at once have continued with the right foot, recovering your point into _seconde_ against his chest and gone right on to his body.

211

Pl. 106.

Now follows a hit in _seconde_ arising in a similar way. The adversary also was in _seconde_; you have moved to engage his sword in _tierce_ on the outside accompanied by the cloak; enticed by the opening you have given he has tried to cut in _mandiritto_. With your sword and cloak in conjunction you have parried with the sword by forming a cross, a much safer method of parrying with these weapons than with the sword and dagger from the certainty you have of defending the lower parts by the cloak hanging from the arm; immediately after the parry you have disengaged in _seconde_ on the inside and hit over his cloak arm,whilst he had lowered it to defend the lower parts; leaving your cloak on his sword you have carried the left foot so far forward that you have pushed his sword into an angle, as shown, an excellent and important result with these arms.

212

<u>Pl. 107</u>.

The next is also a hit in <u>seconde</u> over the adversary's cloak arm. He has made a lunge in <u>tierce</u> on your moving to engage his sword on the inside; he has disengaged on the outside and made this stroke, carrying his cloak to defend his right side; in that time you have changed your hand to <u>seconde</u> and disengaged; resting your cloak on his sword and advancing you have passed through the gap between his arms and thus hit in the chest over the cloak. The hit might have followed from your making a feint of hitting in low <u>tierce</u> between his weapons, and his trying to parry and hit in <u>tierce</u> also; you have changed your hand to <u>seconde</u> and raised your sword so as to avoid his and keep it free, also changing the front of your body and resting your cloak on his sword. Or it might have arisen in another way: being in <u>quarte</u> with your point over the adversary's cloak hand, you have disengaged in the middle with a feint of hitting; in the time of the adversary's,who is deceived by this trick, trying to parry and hit, you have returned your point over his cloak hand, changed to <u>seconde</u> and made this hit.

213

Pl. 108.

The last of the hits with the sword and cloak is a quarte on the left foot against a tierce on the right. You were on the left foot, with the feet close together, and closing distance; your adversary has seen an opening past the faible of your sword on the outside and over your cloak hand, since your sword was in tierce at such an angle that its point was above your cloak hand, on which it was supported for greater security and strength. Not realising the danger he has moved to hit that part in tierce in the straight line, thinking he would exclude your sword. You have changed from tierce to quarte , disengaged your point and driven it on with the left foot; you have lunged with your cloak hand in conjunction and hit in quarte, parrying with the cloak and carrying your right side forward in order to lengthen the reach; this has caused a change of front and lifted your adversary's sword, as shown.Or both may have been in tierce; you have advanced with a feint of hitting in tierce on the outside of his sword, and he has moved to parry and hit in tierce in the straight line at the same time, making sure that his cloak would parry below. You who have moved with cunning, have taken that time, disengaged your point on the inside and in the high lines,and hit with your weapons in union. This has happened because the adversary's cloak was separated from his sword; for if they had been united, your sword could not have passed, since the path would have

been closed. From this you may understand the importance of
the union of the weapons for the securing of good results.

END OF FIRST BOOK.

215

SECOND BOOK,

wherein are explained some principles according to which you can attack the adversary as soon as you have grasped the sword, without waiting for a _time_, principles which are no longer discussed by any professor or writer on the subject.

DISCOURSE ON ATTACKING WITH RESOLUTION.

Thus far we have spoken of the principles with which every professor of arms must be acquainted, though few understand them well or practise them with due exactness. Now we shall treat of some theories, which are not only no longer expounded by other professors, but which they have never considered, or if they have considered have not grasped or understood; they have been put aside as too subtle by the most acute exponents of this art. Desiring perhaps to cover their lack of capacity they have been forced to reject them, basing their reasons on that common maxim, that the student must remain steady in presence and wait a _time_ in order to hit, and that he who attacks without a _time_ will be hit. We allow that is well to know how to await a _time_ and an opportunity to hit, because from that waiting/the understanding of distances, _times_, _counter-time_, and all the tricks which an adversary may employ. Still we maintain that between two opponents steady on their guard there is no advantage, because both are awaiting the same thing, so that the opportunity may come to either; both are awaiting with equal danger, and

and if sometimes one is seen to have obtained an advantage, it
is because he has engaged the other's sword and prevented him
from hitting in his present line; but still the one who has
obtained the advantage waits for a _time_, thinking he cannot hit
before his adversary moves. By such delay it often happens that
one who has won an advantage not only loses it, but his adversary
obtains an advantage over him, a truly inexcusable error that
a man should allow himself to be robbed of what he has won with
such danger. It appears to us that it would have been better,
having the advantage to proceed without waiting, secure that
your adversary's sword cannot hit in its present line, and not
to give him time to consider his danger and form a new plan.

There are others who, when within distance, seek to gain
no advantage, but, seeing that their adversary does not move,
try to make him move by giving him an opening or offering a
time or by an _appel_ or a feint in order to take the _time_ of
his moving; such methods may succeed against men ill-instruct-
ed, but are fatally dangerous against one who understands the
art; for if you consider such a proceeding, it is clear that
the one who offers a _time_ in order to make his adversary move
forgets that his is the first danger, and, although his intention is to offer so small a _time_ that his adversary cannot hit,
still it cannot be so small that the adversary has no chance
of seizing some advantage, from which the first man cannot free
himself without great danger of being hit, moreover he could

be deceived by feints. We do not condemn these theories and
stratagems nor any of the principles already put forward; it
is well to understand them, but they are useless and in-appli-
cable in our case, when we have to find a way of proceeding in
order to be able to hit the adversary immediately after grasping the sword and without remaining steady, in whatever position or guard the adversary may be, whether he offers a _time_
or not, parries or hits, advances or retires; the object is
to hit him inevitably, whatever method he adopts. If our method is followed with all its conditions, you will be incomparably
safer than when waiting. It is true that much skill and art
are needed so to control your adversary that you may be confident of hitting, whatever he does or however much he knows, and
when he has the same weapons as you, even though he is ignorant
of these principles; if your adversary understands these same
principles, matters would be equal. But if your adversary follows the old rules even to perfection, he will always be defeated if you follow our principles, because you will be able to
put him into subjection and free him to do what you want, whether he desires to attack or defend; this will make your proceedings easy, since you will foresee your adversary's intention. In order to explain this truth better we shall treat
first of the advantage of attacking with resolution, and then
of the method of attacking.

218

ON THE ADVANTAGE OF ATTACKING THE ADVERSARY
WITHOUT REMAINING STEADY ON GUARD.

You must consider first that one who is steady on his guard and desires to move will be slower owing to his weight than one who is already in motion. For one who is steady and has both his feet on the ground can move a foot only in two _times_, one in moving it and the other in bringing it to the ground, as we have said elsewhere; but one who is on the move has always one foot in the air, so that without doubt he will have brought it to the ground before his adversary has even lifted his, a matter of great advantage to have finished the movement, while the opponent is beginning. Further the man who remains steady gives his adversary a better chance to estimate his quality and find a way to attack him, than if he approaches without stopping; before he makes up his mind what to do , the opportunity has gone. There is no doubt that _times_ are more readily taken by one on the move than by one who is steady, for the _time_ has passed in moving to take it and it is too late; often a man is hit from such a cause. The disadvantage of the one who remains steady is even greater - for he may be disordered by many kinds of feints, _appels_, and various changes of line, whereas against one on the move feints are impossible, _appels_ only possible by breaking ground and changes of line only occasionally possible; feints and _appels_ are impossible, because he would have arrived before the movement of the feint or the _time_ was finished. In order to advance in the proper manner a threefold

union of the sword, foot and body is needed, and when one of
them is lacking, the method is imperfect; therefore the union
must be observed without rushing with the body or sword.   We
shall now treat of the manner of working with the feet,the body
and the sword together. This is the foundation of the whole
method.

ON THE MANNER OF WORKING WITH THE FEET, SWORD
AND BODY IN ATTACKING THE ADVERSARY WITHOUT A PAUSE.

If you wish to advance against your adversary you must
begin by carrying the feet with ordinary steps, as in walking,
though with somewhat quicker and shorter steps; you must never
lengthen your step except when the point of your sword reaches
his body; your steps must not be violent, for as you must continue until you have reached the adversary's body, any violence
would so disorder you, that you would be unable to lift the
rear foot with the necessary swiftness, and thus by your slowness you would lose your union. Further you should bend the
body forward and make yourself small, so that on approaching
the adversary you can take all the opportunities of defence and
attack with little movement. Your body must not be bent to the
inside nor the outside, except when you are within distance,
when it must be bent to the one side or the other or go straight
according to the movements of your adversary. You must try to

use your sword in such a way, that it is so near your adversary's sword that it appears to be bound to it when it moves,and so that it cannot move without being followed, in brief so that the swords are always united. When the swords are far apart it is a sign that a _time_ has been lost; to approach then would be dangerous, and if you continued you would be hit; in such a case it is better to retire swiftly and return again to the position of advantage. As there are several methods of advancing, some more subtle than others, we shall begin with the one which is first practised, and treat of each one separately in order according to the different principles involved in then.

THE FIRST METHOD IN ATTACKING WITHOUT A PAUSE.

In advancing against an adversary in whatever guard, you must realise his weak and his strong part, the part covered and the part uncovered; you should place your sword in the line which is weaker and more uncovered, beginning with the arm extended and the sword straight in such a way that on reaching his point your point is somewhat higher and stronger, but without moving your point; the nearer you are to his blade the better, taking care not to touch it. Keeping your arm steady you should glide along his blade to his body, without ever leaving the blade, and bring your hilt to the place where your point had begun to penetrate his; you must try to keep his point al-

ways underneath, which may be done with little difficulty if
you are in _tierce_ or _quarte_ directed towards the adversary's
body; if you are in _prime_ or _seconde_, although you cannot control his point above, you can still do so on one side or the
other according as his point is more to the inside or the outside; in this case you should run along his blade, as explained in such a way that as you advance your hilt must approach
the spot where your point was before. This running along his
blade with your hilt must be accompanied by a continued advance,
without withdrawing the arm, thrusting with violence or rushing,
whatever may happen. In brief the method of proceeding is to
be secure that,while your adversary's point is in prime, you
are always stronger, and if he tries to thrust your sword away,
his point will be lifted out of line; when it is lifted your
body, which is in motion, will always pass on before his sword
can return.

If your adversary tries to retire and break ground, you will
not be able to penetrate his point; then it will be convenient
to take the _time_ of his withdrawing in order to force your sword,
and to disengage with the wrist only,without stopping or moving
your arm and making only a small circle with the point; by continuing your advance you will in this way exclude his sword without bringing your own out of line, and if he returns to force
your sword, you will be so far advanced that there will be no
need to disengage, if on the inside, since by simply changing

the hand to _seconde_ and lowering the body you could go on to hit, and would do so before he could push your sword away; if on the outside, you could hit by changing to _seconde_, lowering the body and disengaging your point below, without dropping the hand; this would lead to a hit in the adversary's right side at the moment when he expected to push your sword away. In this manner your body would have passed on the outside without any danger.

This method serves equally when your adversary lets your _forte_ penetrate and then tries to push it away in order to defend himself. It sometimes happens that the adversary tries to push your point away, when it is beginning to penetrate; then it is well to disengage, for, as we have often said, there is no strength in the point. Also it may happen that the adversary disengages, and attacks your point on the other side, leaving the body, and in order to do so before your _forte_ penetrates, he advances his body; in that case, seeing his intention, you should counter-disengage before his sword touches your point, for all disengagements made after the adversary has touched your sword are always more dangerous, since they are made in a bad _time_; the greatest difficulty in the present method consists in this, that you must always be near the adversary's sword and disengage before your sword is found by his; nor must you hold your sword rigidly thinking in that way to offer greater resistance, because it will be found before you disen-

gage. The strength of the sword must be based on your position and not on the force of the arm or wrist. If you follow our method, you will always be prompted to take the opportunity of disengaging or not according to the occasion.

We must consider another case which often arises, that is when the adversary changes his guard and breaks ground,so that you cannot hit in that _time_. Still you must not stop, although there would be no danger in doing so and then returning with the method best suited to meet his change. Yet it is far more expedient that your point, which has already begun to penetrate should follow his point, but only by a movement of the wrist, the arm being kept steady; you should push on and run along his blade to the body. Thus you will deprive your adversary of the power of doing anything; if he tries to make any other change he will be hit during the change, and all because you will be so close to him that he cannot break ground; it follows naturally that you, who are advancing, move more quickly than he, who is retiring. If you wish to stop when your adversary retires and changes his guard, he can always change and break ground, whenever he wishes, so that your proceedings will come to nothing. Therefore you should never stop, if you understand the true method; but if by chance you lose the advantage,then it would be necessary to stop and form a new plan.

This method of working with the arm extended and the sword straight, as explained and as will be illustrated in a plate in

its place, against some guards formed on the same principles requires subtlety of judgment and an understanding of the relative heights of the hand and the point and of the weakness and strength of the method. Since with this method you begin to seek your adversary's sword when out of distance by extending your arm, in order to approach within distance with greater security, it appears to help the adversary by giving him time for consideration and forming his plans. In this respect other forms will be found more expeditious. Still this method is necessary and helps greatly to the understanding of weak and strong parts, the difference between large and small movements, the exactness of the position of the arm, the preservation of an advantage, and of the defence, which should be observed even when hitting. Therefore as a method so important and necessary we have placed it first But subsequently we shall treat of a guard formed high, with which you may attack your adversary resolutely and hit whatever he does or whatever his capacity.

225

Pl. 109.

In this plate we show the method of taking the first advantage in advancing with resolution against your adversary without waiting for a time. If on your approach the adversary offers a time, you should take it or any other opportunity that is presented, following on to the body without stopping. The advantage of this position consists in the fact, that your sword is held above, for two reasons: the first, because it is better to be above than below, the second because the man whose sword is above is quicker to move and form a plan. In this superior position you should continue to the adversary's body, running along his blade, and in advancing bring the hilt of your sword to the spot where your point was first, without moving your point away from his blade, until you hit. If his sword had chanced to be in tierce at an angle or in quarte, you should have begun in this way, but without running along his blade; you should attack with your blade in a straight line from its point to your body and hit in the gap made by that angle both on the inside and on the outside. Other plates will follow, which will show what may arise from this advantage, but they will not be very many for the sake of brevity; the principle and most necessary ones will be included, from which the remainder may be understood. Also, some remarks will be added in the texts.

Pl.110.

226

The hit in _quarte_ here represented has followed from this first advantage. You have run along the adversary's blade,bringing your hilt to the spot where your point began, and reached the part shown, because your adversary was too slow in moving and unable to defend himself or do anything but draw back his body, without success. After getting control over his point, with the right foot forward, you have passed with the left foot and then with the right, and so pursued your victory right to his body. You would have done the same, if you had begun on the outside, the only difference being that your sword would have been in _tierce_ instead of _quarte_, while your adversary's sword, instead of being driven upwards by the force of your sword, as in the plate, would have been driven downwards by the _tierce_, and your point would have hit below your own hilt; in that position it would have been stronger and would have given more protection to your lower parts.

227

Pl.111.

The next hit has also arisen from the same initial advantage, and in this manner: in advancing you were running along your adversary's blade, having already brought the left foot in front; he has tried to parry by forcing your sword and drawing back his body; feeling the pressure, you have changed your hand to _seconde_ and given way to his sword, which has gone to the outside through meeting no resistance, and all the further out of line because you have lowered your body. You have kept the hilt of your sword at the same height against his _faible._ The angle formed by the _seconde_ has carried your point to a hit. If you had attacked on the outside in _tierce_ the result would have been the same. After bringing the left foot forward, if the adversary had tried to parry, you would have again changed your hand to _seconde_, put the point under his right arm, keeping the hilt at the same height, lowered your body, followed on with the right foot and thus made the same hit. But if you had begun on the inside and the adversary had begun to parry by breaking ground,as he could you should then have disengaged with the wrist on the outside in _tierce_, and gone on until you reached his body. If again he had proceeded to parry, as he might, without breaking ground, you should hit in _seconde_ below. But if he parried at the point of your sword, if you had begun on the outside, you should disengage in _quarte_ on the inside, and if he again parried and drew back, you should

turn the hand again and hit in _seconde_. If at the beginning
the adversary disengaged in order to hit, you should simply
go on in the straight line in _tierce_ or in _quarte_, according to
whether you were on the outside or the inside, and you would
hit in the _time_ of his disengage. When the adversary disengaged,
if he had not advanced but defended himself by breaking ground
whether on the inside or the outside, you would have been sure
of hitting in that second _time_. It might happen that at the
beginning of the movement he disengaged and broke ground with
the idea of engaging your _faible_; in that case you should
counter-disengage, before he touched your point, and follow
on in the straight line, so that if he makes a double disengagement, as he might, you could defend with little movement
and without disorder. If after his first disengagement he returns to the parry, you should hit above, as we have explained
since all he could do would be to use his left hand, which
would cause only a slight disturbance or might even give you an
advantage. All these rules apply against a _tierce_ or a low
quarte, whether straight or at an angle. Afterwards we shall
treat of the _prime_ and _seconde_, but we have put these first,
as being more usual.

229

PLATE 112.

The next plate shows how to take advantage of an opponent, who stands with a low guard and with his upper parts held back in order to protect them, since they are more exposed. You have the advantage of the higher position, and let your point run along the adversary's blade, keeping it above his blade and gradually raising it; as the point is raised so the hilt is lowered, so that when your hilt reaches his point it is in the position of the point, as seen at present; thus you run along the blade until you reach his body. If the adversary disengages, you will make no change, nor even bring your hilt any lower, but hit without any other defensive movement, realising that his point is excluded by reason of the shortness of his sword and the length of time of his disengagement, since he is below, and wishes to reach the upper line. Finally the point of your sword achieves the same result, as if he did not disengage. In order that the position may be better understood, in the next plate we shall show the hit which follows from this advantage, from which you will understand another result which may arise.

230

PLATE 113.

Here then is the hit in _tierce_ against another _tierce_, which has followed from the advantage seen in the preceeding plate. Both combatants had their points low and you were above; you have followed on,running along the adversary's blade, and raising your point and lowering your hilt as you advanced, have made the present hit and continued right to his body. The other result which we said would be illustrated is this; you may have found your hilt inside his sword, and as you advanced he may have disengaged in order to free his sword; but you would have prevented him by keeping his sword below, because your point was in line; all he could do was to try to push your point out of line, and in that he would have failed. If he had again tried to disengage, he would have effected nothing, because you would have made the same hit by a mere change of the hand towards _quarte._

231

PLATE 114.

The next position is also a _tierce_, but differing from the preceeding one in that the advantage has been acquired against a _seconde._ You have begun at a distance to hold your sword in the manner shown so that on reaching the adversary's point you have acquired the advantage without any movement of the hand or point. With this slight advantage in _tierce_ you can go on, following his blade, without however touching it, and bringing your hilt to the present position of your point; as you advance, you should turn your hand in such a way that you are in a guard of _quarte_ when your point makes the hit. As the point is now higher than the hand, it will then be lower; you will hit in the chest, keeping your hilt in the defensive position.

In the next plate we shall illustrate the hit which you may make in this position in _tierce_ against an adversary who makes no change.

232

PLATE 115.

Here follows the hit referred to at the end of the last
discourse, made by a _quarte_, originally a _tierce_, against a
seconde and perhaps following from the advantage described.
You have continued along the adversary's blade as explained,
until you have reached this position; by bringing forward
the left foot and then the right, maintaining your defence
with the hilt, which has approached his sword, without stopping you will go right on to his body for greater security.
Or it may be that you have acquired the advantage on getting
within distance, and the adversary has disengaged in order
to free his sword, and withdrawn for his protection. You, being already on the move, have arrived so quickly with a
counter-disengage, that he has been able to form no plans,
and all because of the advantage of the continued advance. If
you had been slower and not advanced in the _time_ of his disengage, you would not have reached, and thus would have given
him an opportunity to parry and hit, before you arrived.

233

PLATE 116.

The next plate shows a hit in a guard of _prime_ on the outside under the sword against another _seconde._ You have begun with your sword extended and have advanced to engage, holding your   point against his sword on the inside; you have reached his point with the right foot and followed with the left. The adversary has tried to disengage in order to hit in that _time_ over your sword on the outside, but by bringing forward the right foot in the same _time_ and bending your body you have carried your point under his sword arm and hit by turning the hand from _tierce,_ into a guard of _prime_. In this way you have guarded against the danger of his making of feint of disengaging, and have excluded his sword in such a way that he could not bring his point into presence. The result is due to the fact that the adversary has allowed your sword to advance too far before disengaging; if he had disengaged on your first approaching his point, you could only have counter-disengaged and hit in _quarte_.

234

THE SECOND METHOD OF ATTACKING THE ADVERSARY
WITHOUT A PAUSE.

The most necessary guard in this method is a high guard
of tierce, formed with the whole chest facing the front, the
points of both feet towards the adversary, the body bent forward, the sword hand near the face, and the point of the sword
suspended in the air and advanced, but not so far advanced that
it may be engaged by the adversary before he is within close
distance. With this method and this guard you must advance
against your adversary with natural steps and towards the outside, until your body is so far outside his sword, that your
sword too is in that part without any movement of the hand.
The sword must be kept motionless, and as you approach the
body lowered, so that the point also will be brought lower,
and so far that when your hilt has reached his point your
point will be in line. On moving to hit you should not lunge
but approach your body to the adversary's. If he changes his
front or moves his point to prevent its being left out of line,
you should take that time, place your sword on the inside,
without extending the arm and keeping both sides equally advanced, and bend the body so far forward that your point comes
into presence. You must avoid dropping the arm or hand, since
the hand must face his point until the whole body has passed,
whether on the inside or on the outside. You should always

begin on the outside, whatever your adversary's guard. Although when within distance you should be prevented from passing outside by your adversary's changing his line or front,
you should still, having reached his sword, go on, resolutely
straight to the body. Even if your adversary's point is low
and pointed towards the ground, you must still lower yourself
until the _forte_ of your sword excludes his sword on the one
side or the other, but without moving the arm. But if his
sword is low and on the inside, as you lower the body to reach
your position you must carry the right side back in order to
bring the body out of line, so that if the adversary disengages he could not reach your body while you are lowering it,
and in order that you may go straight on to hit without defending.

With this guard we must warn you against disengaging, except when the adversary raises his sword in order to engage
yours, when it would be convenient; but it should be done
without movement of the sword or arm, and with a slight turn
of the body only, keeping back the right side; in this way
your sword hand will be drawn back, so that the adversary will
not reach. At the same time you should carry your left foot
across, so that your sword will be brought outside without
being moved, and your body will be protected from the adversary's hit, whilst making this turn. If on the other hand

you changed your front and did not move to the outside, you would expose yourself on the inside. Moreover as you move to the outside, your adversary might raise his sword to parry, which would be an opportune moment to dash under his sword, turning your hand and keeping it at the same height.

When this guard is well formed, the only weak points are the inside and the face, but as the hilt is close to the face, it can readily be defended. Also, as the lower parts with this guard are distant, they are in no danger, except from feints which the adversary may make; if he feints on the inside, in the time of your swords dropping to parry, he may lower himself and hit below by passing. But if you are careful not to move your arm in parrying feints, but rather to bring the body down to that part, your hilt will parry of itself without moving the sword; if your adversary tries to pass, he will be hit, because your point can reach any part down to the ground, before he has passed. On the other hand if your arm falls or is extended, you will then certainly be hit. As to cuts, with this guard they can do little harm, for they can attack only half the head, and your hilt is very near that part. Cuts cannot reach below, nor can they strike the sword in order to disorder it. Therefore this method is very good in attacking without a pause, and so much the better as the sword is secure from being engaged by the adversary. But if you pause it is not so good, since you may easily be

thrown into disorder by the adversary, unless you resolve to
see your left hand.

PLATE 117.

This plate represents the guard in _tierce_ discussed
above, and is to be used with this second method. It is formed high,out of line, with the hilt close to the face, the body
bent and the feet close together, all with the object of keeping the sword free from being engaged except with certain danger to the adversary of being hit. Since the chest is facing
the adversary, the guard can be disturbed only on the inside;
in order to facilitate the defence you advance against the
adversary by moving in a circle towards the outside, so that
when within distance your sword and body are outside his sword;
if they were not outside it would be due to some change made
by the adversary. The lowering of the body with this guard
varies more or less according to the relative height of the
adversary's sword. The head is near the hilt for greater security and strength, and for greater speed in advancing.

238

PLATE 118.

This high guard of _tierce_ leads to the advantage shown
in this plate. As you approach the adversary, you direct your
point, and lower the body, keeping the hand and arm steady,
until the point is in line. You have controlled the adversary's sword and will maintain that position, when your point
hits. If, when you come into line, the adversary disengages
it order to hit on the inside of your guard, you will have
penetrated with your body half the length of his blade, which
deprives him of the power of turning out of line or passing.
The advanced position of the guard facilitates your passing
below, if he attacks your sword. in order that that result
may be better understood it will be illustrated in its place.

239

PLATE 119.

From the advantage acquired by this high guard has arisen the following hit; you have reached the adversary's point with your _forte_, checked your hand in that position, lowered your point, directing it against the adversary, and by carrying your head forward and below your own hand have continued your advance and hit in the throat, when his sword was high; if it had been lower, you could have hit lower by dropping your body and hand in proportion. If he had tried to parry by raising his sword, you would have disengaged below, and by turning your hand to _seconde_ and keeping it in the same place would still have certainly hit.

240

PLATE 120.

The next is a hit in _quarte_ against an opponent who has
lunged in _quarte_. You had engaged your adversary¦s _faible_
on the outside with the high guard shown in the previous
plates; he has tried to disengage and lunge in _quarte_ on
the inside in the path which appeared to him to be uncovered. But with your arm high and held back you have pressed
on his _faible_ with a guard of _quarte_, and by carrying the
right foot forward and somewhat out of the straight line and
bending the body over it, you have made a hit in the throat.
By continuing with the left foot you would have passed right
to his body. If the adversary had tried to turn with either
foot, he would have been all the weaker, and therefore you
would have hit in the straight line or in the back.

241

Pl. 121.

The next is a hit in _seconde_ made by passing underneath
with the body on the outside, against a guard turned somewhat
towards _quarte_. You were in a high _tierce_, with the hilt near
your face and advancing towards the outside; the adversary in
tierce in a straight line, seeing you coming, raised his point
to impede your sword and cover himself above. Being already
within distance you have left your hand at that height, turned
to _seconde_, before he could reach your sword, and by bending
your body as far forward as possible have passed, With the head
so far advanced as to penetrate the whole length of his blade;
thus you have hit, and the adversary has been unable to defend,
nor has had time to bring his sword back into line or to turn
his body.

242

Pl. 122.

This plate illustrates the exclusion of an opponent's sword in low _tierce_. You were in high _tierce_ and have lowered your body as you advanced, so that on arriving within distance you were so low that you have controlled the adversary's sword without moving your arm. While lowering your body you have carried the right side back and the left forward and so balanced the body on the feet and knees, that you have been able to get so low with great swiftness. The position of your body is such that even if your adversary had disengaged above, you would not have checked your advance or made any defence since there was nothing to defend in that line. Whether he had disengaged or not, or whatever he had chosen to do, he would not have prevented your advance, with your sword, feet and body working in union. The change in the position of the body after reaching this point will be shown in the next plate.

243

Pl. 123.

From the advantage shown in the previous plate with a high
tierce has followed the present hit in quarte on the left foot
against a low tierce. After winning your position of advantage
without extending your arm, you have placed your hilt against
the adversary's sword in the position where your point was originally and by running along close to his blade and directing
your point into line you have made the hit in quarte, carrying
the left foot forward. You would equally have made the hit,if
the adversary had disengaged by keeping your hand in tierce;if
he had tried to raise his point in order to hit above, by simply raising your hand to quarte you would have made the same
hit and passed with equal security.

244

THE THIRD METHOD OF ATTACKING THE ADVERSARY WITHOUT
A PAUSE.

The first method which we discussed on this subject of attacking with resolution is good, because you begin to acquire the advantage so far out of distance, that the adversary cannot hit. Yet it appears that the danger is revealed to the adversary too soon, so that he has good opportunity to change his line in order to disorder you, and ample time in which to employ various devices for his protection. The second method also is good, since it forms a secure guard with only one exposed part and that part so near the sword hand that it cannot be reached without passing your <u>forte.</u> With this guard also your sword, as we have shown, is kept so free, that few disengagements are needed. If it were not in other respects so restricted,and you were not under the constraint of keeping your own steady it would be better than the first. Nevertheless considering the imperfections of these two methods,and particularly that defending oneself when the adversary cannot attack is a loss of time and a disadvantage,since it reveals your intentions to him and gives him a chance of finding a remedy,we have sought for another way of proceeding, a third method,which reveals nothing to the adversary until his body is in danger.This method when properly executed,will hit with such swiftness that the adversary not only has no time for so many changes, but can barely parry the first onslaught.

The foundation of this method is the certainty that the

adversary cannot hit before you are within distance; therefore
there is no necessity to defend or to hold your sword steady in
any position. You should advance towards the outside, until
your feet are within distance; it is of no importance which
foot is first. The time to carry the forte to the adversary's
faible is when lifting the foot to bring it within distance, in
order to exclude his sword without stopping; you should run
along his blade in order to hit with your sword, feet and body
in union and without rushing; for if he should then break
ground he would have time not only to parry but to hit also.By
advancing in union you can change in time, as you should do if
on the inside when he parries; you should in that case change
from tierce to seconde, lower the body and continue your advance when you will hit at the moment of his attempted parry;
but in turning from tierce to seconde you must drop your point
under his arm, keeping the hand in the same place and bend the
body so as to hit in the right side. If your adversary has
succeeded in parrying by breaking ground after you have engaged
his sword and advanced to hit, he can no longer bring his point
into line as for example he could have done, if you had stopped
and made an interval between engaging his sword and advancing,
for your plan would have been too slow. Similarly if you had
rushed your body or sword forward or hurried your steps,you
would have been at a disadvantage, since you could not have
turned a second plan, but rather would have been in danger of
being hit.

246

You should adopt the same method of advancing with resolution if your adversary on your first approach to engage his sword
parries without breaking ground, since before he forced your
sword you could hit and pass. But if when making this parry he
breaks ground, it is then better to disengage, before he touches
your sword; here is the difficulty, because if you move your
sword on first seeking his, you cannot disengage in time. Therefore you must advance in such a way that the movement of disengaging shall not be opposite to your other movement; if by accident your hand fell, you could not lift it again in time, if
your adversary advanced to meet your sword. But if your point
is carried with such ease that you can abandon your first plan
and adopt another according to the occasion and with the necessary skill the method will be very deceptive, since, when within
distance, you engage your adversary's sword and while he expects
to meet and resist your sword you disengage and advance the other
foot, so that he can no longer return into line nor do anything
but hit below by a half-disengagement; in that case you have
only a small movement of the point to make and to lower the body
to the line in which his sword is directed; you will continue
on your course, exclude his sword and certainly hit. But if the
adversary, while you are attacking his sword disengages or advances, rather than breaks ground, he will be hit before he has
finished the disengage. If he disengages and breaks ground in
order to find your faible again, then you should counter-disengage and advance, when you will hit at the same time; this will

be easier and shorter than seeking his sword and disengaging,
before he touches your sword. If the adversary changes his
guard, when he breaks ground, raising or lowering his point or
withdrawing it, in every case you should continue your advance
and again seek his sword as soon as you are within distance,
but in such a manner that in whatever way he tries to hit, you
can keep on your course, parrying and hitting together. From
the position and the distance between your adversary and yourself you will understand what he can do in defence and attack,
how he can disturn and impede your sword and how to guard against
it. For if you do not foresee what may heppen, the opportunity
passes so quickly that there is no time to form a plan.

Of the things that the adversary may do by retreating or
withdrawing we do not consider it necessary to treat, because
they do no harm, still it is well to know them and be prepared
for everything. Those similarly who cut against this method
of advancing we can disregard, merely saying that, if your adversary cuts in the _time_ of your advancing to engage his sword,
he will be hit before he has half finished his cut. If he cuts
while withdrawing you can follow him, covering yourself and
proceed to hit; if perhaps you fail to reach and you have parried on the inside in _quarte_ you can change to _seconde_ and hit
in the same line, where his sword will be unable to parry; if
you have parried on the outside in _tierce_, again you can change
to _seconde_ below and still hit in that line. If you do not
wish to parry, you can let the cut pass, and immediately ad-

vance, not to hit at that moment, but in the _time_ of his raising his sword or recovering or making another cut. This is
better than parrying. An understanding of this third method
is better than that of the two first methods; but you must
have a good knowledge of distances,and without that you will
get no profit from any of the methods we have described, least
of all this last method, with which you advance without pause
and without holding your weapons in any particular position.
In brief you must realise when the foot carries the body into
danger in order to secure yourself. When you have well practised this method you will be able to grasp another method,
which we shall explain next.

249

Pl. 124.

Here you will see a <u>tierce</u> on the left foot, which has acquired the control of the opponent's sword on the outside, the opponent also being in <u>tierce</u>. With this advantage with the same guard you will continue right to the adversary's body. You will succeed because you have approached within distance without forming a guard,and when your foot came within danger you have covered yourself from your adversary's sword without touching it; you will advance without pause right to his body, taking a <u>time</u> according to the opportunity; if he offers no <u>time</u> you will follow along his sword, and continue as you have begun, preserving your union.

250

Pl. 125.

From the advantage won over the adversary's sword, as shown in the preceeding discourse, has followed this hit in _quarte_ against another _quarte_. Seeing you advancing in order to control his sword, the adversary has taken that _time_ and disengaged in _quarte_, turning his body with the left foot, in order to hit you in the chest; but you have advanced in union and with little movement of the sword and by merely changing your hand from _tierce_ to _quarte_ and continuing your advance, you have hit him in the throat in the same _time_; also you have driven his sword out of line, because the position of your sword was stronger, and because of the weakness of his position in turning his body and because your arm is stronger than it would be if extended.

251

<u>Pl. 126</u>.

The second hit below the sword on the outside has also
followed from the same initial advantage. You have moved to
engage the adversary's sword on the outside at the moment when
your foot came within distance; the adversary, who was in
<u>tierce</u>, has taken that <u>time</u>, changed to <u>quarte</u> and dropped his
point in order to free it and hit in the right side below,turn-
ing his foot in order to carry his body out of the line of your
point; seeing his object you have checked your hand in the position where it was, turned to seconde, bending the body forward
brought your point underneath and excluded his sword,before it
came into line . In this way you have made a hit in the side,
following on to his body without stopping.

252

Pl. 127.

The next plate represents a control acquired by a <u>tierce</u> against a <u>seconde</u> in this manner; without forming any position you have advanced and placed your sword in <u>tierce</u> against the adversary's sword at the moment when your foot brought you within distance, his sword being in <u>seconde</u>; without touching his sword, you have covered yourself and prevented his hitting in his present line. With this control you can go on to hit in <u>quarte</u>, carrying your hilt to the present position of your point against his sword; if he disengages as you advance to get the control, you can continue your advance by a counter-disengage in <u>quarte</u> and hit just the same. If it chanced that you were not so far advanced, that you could avoid his point by lowering the body, then by merely changing your hand to <u>seconde</u> you would hit below in the right side, and let his sword pass in vain above.

253

Pl. 128.

From the same advantage acquired by a <u>tierce</u> against a <u>seconde</u> explained in the last discourse, has followed the hit now illustrated.   Having controlled the adversary's sword with your <u>tierce</u> and seeing that he made no movement, realising also that you were defended in that line without need to touch his sword, you have passed on with the left foot and maintaining your defence have made a hit in <u>quarte</u> in the throat;   you have kept your hilt against his sword and bent your body as far forward as possible.   You see too that the heel of your right foot is raised, which shows how the pass will be continued right to his body.

254

Pl. 129.

This hit also has arisen in a similar way. With the guard of _tierce_ against the adversary's sword you have covered yourself and acquired the advantage. The adversary has taken that _time_ and, lowering his body and point, has carried his right foot forward in order to hit below the sword, while you were trying to find his sword. You with a guard of _tierce_ have begun your approach with little movement of the sword and without hurling it forward; seeing the adversary's plan you have abandoned your first policy and adopted another; by lowering your body and point in such a way as to leave your hilt against his sword and exclude it, you have hit in the chest as he was advancing; if he had changed from _seconde_ to _tierce_ to defend himself and thrust your sword away, he would still have been hit, because you would have changed from _tierce_ to _seconde_ and by lowering the body and continuing your advance you would still have hit in the chest, without his being able to parry or recover his point into line, because you would have passed before he started. Similarly if he had disengaged to hit in _seconde_ on the outside above the sword, by changing to _seconde_ and lowering your body you would have hit below his sword. All these methods proceed excellently because of the advantage of the continued advance and because that guard of _tierce_ has provoked the adversary to move.

255

THE FOURTH METHOD OF ATTACKING THE ADVERSARY WITHOUT A PAUSE.

The fourth method, which we are now about to explain, also has no fixed position for the sword in approaching the adversary Whereas in the preceeding method you advanced against his _faible_ and tried to be on the outside on arriving within distance, with this method you proceed in the opposite way and with greater subtlety so that the adversary cannot know your intention. In that method you tried to get to the outside, if possible, in this you advance with your chest facing his point, so that it appears that you intend to rush in; thus the adversary can on-ly remain in the straight line in order to hit your body, which is advancing uncovered. But your intention is, when your foot is about to enter within distance, to carry it outwards on either side according to the occasion; if you enter within distance with the right foot, you carry it out to the right, if with the left to the left; thus one of your feet remains in the straight line and the other outside; the body is always bent over the foot on the outside, which brings the body also out of line,and uncovers the adversary's body. If he tries to hit at that time, your sword is not far from his and will easily defend. But if he does not move, you should then advance in that line to which the foot has crossed, excluding his sword, in order to hit with the next step, which should be short and the advance continued for greater swiftness. If it happens that the adversary directs his point towards your body which is being bent over the foot

to the outside, you should carry the other foot forward, for it will be already raised, and bring the weight of the body on to it and thus out of line on the other side; this will exclude his sword, and you may proceed to hit.

This is a method to be followed when the adversary holds his sword higher than the middle of his trunk, more or less; but if his point is directed towards the knee or lower, then you should advance against his point and,on the instant of your foot coming within distance, cover his sword so that he cannot raise it; but you must take care in covering his sword not to let your point fall below his blade, since he would hit without your being able to parry, and you would be forced to counter-disengage. But if you hold your sword in the proper manner,you would hit in the _time_ of his disengaging, without making any movement to defend, if the adversary were on the outside; if he were on the inside you should make a slight turn towards _quarte_ without changing the hand entirely and taking care not to bring your hilt too low, so that he could hit in the angle formed from the hand to the point, as shown in the plate; for in that case you could parry only with difficulty, and even if you par-ried would be in danger of a hit in another part owing to the large movement you would make. But if you advance with your sword in the exact position and accompanied by the foot and body, your success will be complete. Therefore this method is better than those already described. Afterwards we shall treat of another method, similar, but involving more subtle princi-

ples and requiring greater judgment, because you are brought
into greater danger; on the other hand you make your hit more
easily, and if you thoroughly understand it's principles, you
will advance with safety and hit without impediment; for the
method is very deceptive, more so than the others of which we
have spoken.

Pl. 130.

With the _quarte_ here shown the right foot has been carried
away and the weight of the body brought on to it; the sword
has remained in the straight line below the adversary's in
tierce. You have begun at a distance and approached with your
chest facing the adversary's point, until your left foot was almost within distance, when you have carried the right foot away
and brought the weight of the body on to it with the object of
bringing the body out of the line of his sword and of being
able to put your sword in the position you judge to be best.
Since your sword is exactly below his, the adversary has been
unable to engage it with ease, and has been left in doubt. If
he has not moved on your carrying your right foot out,you could
place your sword in the line uncovered close to his sword, exclude his sword, and go on to hit without touching by advancing
the left foot on the inside. If the adversary has followed with
his point the line of your body now inclined on the right foot,
you would have brought your body on to the left foot and out

of line on the other side; on hit sword moving you would have put your sword into the line uncovered, for since your sword was just under his line, and since he has followed your body with his point, your sword would be left in that line, so that by merely thrusting it forward in the line where his sword was you would exclude his sword, and with all the greater ease because of the movement of your body, now on the left foot, a movement quicker than that of the hand. By following on with the foot you would have passed with great swiftness, leaving his sword on one side or the other according to the circumstances.

These movements must all be made continuously, without stopping although in the plate you appear to be awaiting a _time_ with your feet apart, that is done to show the position of the foot, body, arm and sword; but actually the movement must be executed swiftly and without interval. For if the adversary does not follow the line of your body, you will at once advance, close the path of his sword and continue. But if he follows your first movement with his point, your body goes to the other side and leaves his sword out of line, so that it cannot return into presence. Whether you make the side step to the right or the left, you must still leave your sword and hand in the line of the adversary's point in order to make your defence easier, if he tries to hit during your movement. You will succeed well with this method, if you proceed in the correct manner, remembering that you must know how to reach the petition shown with-

out moving the arm or sword; the sword must be carried forward
by the body, or the method will be dangerous.

Pl. 131.

By carrying the right foot outwards as shown, this tierce
has acquired an advantage against another tierce. On your carrying your right foot outwards, the adversary has not moved;
your body is out of line and has uncovered his; you have immediately lifted your left foot, excluded his sword out of line
so that it can return only by a disengage, and you would have
hit before that movement was completed; and carried the left
foot towards the line of his point. If the adversary had disengaged, you would have carried your right foot to the line of
the left and hit in tierce without any other movement than that
of extending the arm, rendering his disengage useless. If he
had not disengaged then without carrying the right foot into
the line of the left, you would have hit in quarte and followed on to his body without touching his sword. If he had rushed in in order to defend himself you would have to rest satis-
fied with covering yourself against his attack.

260

Pl. 132.

This hit in quarte has followed from the advantage of having excluded the adversary's sword on the inside, as shown in the preceding plate, and from the fact that his sword, which was then in tierce in a straight line, is now in tierce at an angle. As soon as you were out of presence, you had placed your sword in the line uncovered, thrust past the faible of his sword without touching it and brought your hilt to his sword; your hand in its advance has driven his sword into an angle, and the further the advance the greater the angle, so that he has been unable to do anything but parry. All this is because with this method, when you are ready to hit, you have penetrated at that moment so far forward that the adversary is unable to form any plan except that of retiring and parrying. And also, even if you change your line you still hit without the adversary's having any defence. Of such importance is the advantage of the line, the position of the feet and the distance.

261

Pl. 133.

This plate represents a <u>tierce</u> below another <u>tierce.</u> As explained you have advanced from a distance, and when nearly within distance, your right foot being in front, you have carried off the left foot and brought the weight of the body onto it in order to bring it out of the line of the adversary's sword. You are holding your sword slightly below his in a straight line in order to be ready for the defence and to be able to place it where needed with little movement. Since the adversary has not moved his point, with this <u>tierce</u> you will carry the right foot to the line of the left, thus bringing yourself entirely out of line while his body will be exposed on the outside; in the same time you will place your sword in the line uncovered and hit close to his sword without touching it. But if the adversary follows your body with his point, when you move the foot away, the body which is on the left foot will be carried onto the right foot and thus out of line on the other side; by advanceing the left foot and putting your sword in <u>quarte</u> inside his, you will hit without touching his sword. The result will be excellent, for in this position you will always have one foot out of the line of the adversary's sword and can advance on that side, if he has not followed you with his point; if he has followed you, you can bring the body onto the other foot, put your sword in that line, preventing his sword from returning, and hit in the part he has uncovered.

262

Pl. 134.

The advantage acquired by the <u>tierce</u> here represented
against another <u>tierce</u> has followed from your carrying your
left foot outwards. After advancing with your chest facing the
adversary's point in the manner explained and after carrying
your left foot outwards, you have immediately lifted the right
foot, so that your body has come out of line, and at the same
moment have placed your sword close to the adversary's to prevent his attacking; your intention is to go on to his body with
that <u>tierce</u> without touching his sword; if he should try to
find your sword in order to thrust it away, you would be confident of resisting his sword and making the hit, if you had not
touched his sword in the first place. Also if he should parry
and break ground, you would attack below in <u>seconde</u>, before he
touched your sword. If he should disengage on your approach in
<u>tierce</u>, you would change from <u>tierce</u> to <u>quarte</u> and hit by car-
rying the left foot straight forward from its present position,
since his sword would be weaker and your arm would make a
smaller movement.

263

Pl. 135.

From the advantage explained in the preceeding discourse
has followed the hit in _tierce_ now shown. You have carried
your body out of line and placed your sword in the line you
saw exposed near the adversary's sword; keeping yourself covered you have hit by advancing the right foot and following
with the left, since the adversary has formed no plan when you
carried the body away and therefore has not and could not have
parried except by breaking ground; in that case you could have
attacked in _seconde_ below by bending the body and at the same
time penetrating his point with your head before he couldbring
his point into line. If he had tried to defend himself against
the _tierce_ and to hit by disengaging you would still have hit
by turning from _tierce_ to _quarte_ before his disengage was completed; he would have been unable to do anything else because
of your advanced position, as we have said elsewhere.   When
you decide to hit the adversary has only one resource, that is
to break ground whereas you have many.

264

<u>Pl. 136.</u>

From your passing out of presence with the left foot has
followed this next hit. When your foot came to the ground,and
the adversary followed your body with his point, you placed
your sword in the part exposed on the inside near his sword
and thus prevented his sword from returning into line; you
then continued with the left foot right to his body. If he had
tried to parry he could have done so only by retiring and carrying his point out of line; further he could not have succeeded since he was facing your point, so that you could easily
have changed from <u>quarte</u> to <u>seconde</u>, bending the body, which
is now inclined to the right, over to the left to the position
of the <u>seconde</u> on the inside, but somewhat further out and low.

265

Pl. 137.

The next plate represents a hit in _tierce_ against a _seconde_ intended to hit below the sword. You have carried your left foot away, and advanced the right, putting your sword close to the adversary's in order to exclude it. He has taken that _time_ turned his hand from _tierce_ to _seconde_, lowering his body and point in order to hit below. You have not touched his sword, but simply covered yourself; you have dropped your point at the same time, still in _tierce_ and carried your left foot forward bending the body; you have checked your hand in order to remain at his _faible_ and inclined your point upwards, in order to give it more strength above; in this way you have stopped his sword and hit in _tierce_; this _tierce_ has penetrated all the more because it has been met by the adversary's, whose point has been lowered in his efforts to defend himself against the approaching danger but he has been deceived and encountered your forte.

266

<u>Pl. 138.</u>

Now follows a <u>quarte</u> which is below a <u>seconde</u> and with the
left shoulder more advanced than the right. You have carried
away the left foot, lifted the other and brought it into the
same line in order to expose the adversary on the outside. Although the adversary's hand is so high, yet the whole of his
head is exposed above in the line from the middle of his blade
to the point,and thus the line in which this <u>quarte</u> will hit
is seen. If when you carried away the left foot, the adversary
had followed with his point in order to remain in line, you
would have brought your point to the inside in <u>quarte</u> and hit
in that <u>time</u>, without taking your sword away from the defence
and without touching his sword. If he had not moved,with that
<u>quarte</u> you would have attacked on the outside and hit above,as
will be more clearly shown in the next plate.

-267-

Pl. 139.

This hit has followed from the position of advantage of
the preceeding _quarte_ below a _seconde_. You had brought your
body out of presence, and seeing the part uncovered towards
the adversary's head on the outside, have at once placed your
sword in that line, extended your foot and arm, and by running
along his blade forced it down, as shown, for the _quarte_ is
very strong in that line and the _seconde_ on the other hand very
weak. Even if he had tried to change to _quarte_ he would have
done no good, for by simply lowering your point towards his
right side you would have hit at the moment of his advancing,
before he had finished turning his body and hand. If, when
you had gone out of presence, he had followed in order to maintain his point in line, you would have taken that _time_, your
hand being already in _quarte_, and hit in _quarte_ on the inside.

268

Pl. 140.

In this plate you are seen to have acquired an advantage
in _tierce_ over another low _tierce_. Coming from a distance
without forming any guard, you have carried your body and
sword in such a manner, that on arriving within distance they
were in the position shown. If your adversary had tried to
hit on one side or the other, as you came within distance, he
would have effected nothing, but you with your advantage would
have been better able to hit, since in approaching within distance you have kept your feet, body and sword in control, in
order to be ready for any opportunity. If your adversary does
not move you will go on to hit close to his sword and the
line of his arm in order to preserve your defence. If he disengages in order to hit on the outside, you will hit in _tierce_;
if he does not move, you will hit in _quarte_ in order to defend
the inside line.

269

Pl. 141.

This hit in _quarte_ has followed from the advantage acquired by the preceeding _tierce_ on the left foot against a low _tierce_. It has followed because you have seized the advantage and continued close to the adversary's sword; you have so defended yourself, that if he had disengaged on the outside, you would still have hit without making any movement of defence in tierce or any other change; he could have saved himself only by breaking distance and raising his sword to the defence on the one side or the other; this would necessarily have brought his point out of line and given you a good opportunity to hit in _seconde_ on the inside or below, according to the direction in which he had moved to parry. Your body would have been so far on the outside, that you could have kept your right shoulder exactly opposite his right shoulder.

270

THE FIFTH METHOD OF ATTACKING THE ADVERSARY WITHOUT A PAUSE.

Now we shall treat of another way of attacking the adversary, which is more subtle than the others. If you can safely reach the required position, you will hit without danger. On approaching, whatever your adversary's guard you should gradually bring your sword towards the position where you intend to place it, so that on arriving within distance your sword reaches, the exact position desired. As we have said several times, the sword must be placed against the weakest part of the adversary's sword; this is so with the present method, until you are, within distance; but although elsewhere we have taught you to put your point against his point, yet with the present method you must advance so far as to bring your point against his hilt, but without letting his hilt penetrate beyond your point, though near it; your point must be in line with his hilt, neither above nor below, but to one side according to the position adopted by the adversary. Your point should be inclined rather downwards than upwards, for two reasons, the first, that you may be better able to disengage if, if necessary, the second, so that the adversary can find it only by lowering his hilt, which would give you a _time_ to hit, as you are already on the move and your point is very near the adversary.

On reaching this position, if your adversary is in _tierce_ or _quarte_, you must hold your sword in a straight line from your wrist to the point and your arm so far advanced that you are sure of defending any possible stroke with little movement,

either when you reach your position, or at any time. In brief
your sword and body must be in such a position that your _forte_
can defend with little motion . But if your adversary is in
a guard of _prime_ or _seconde_, you must then place your sword
exactly in the line of his hand, but below it, and hold your
sword in such a way that your hand forms no angle, and if he
tries to hit, you could parry with the same guard and hit at
the same time on the outside over his sword, carrying the foot
in that direction to shorten the movement, and to give yourself
better cover and greater strength. If your adversary does nothing when you have your sword in the exact position, you should
raise the point above the line of his hand and go on to the
body and the nearest part exposed, covering yourself with your
hilt in the line, where his sword might come, and supporting
the stroke by the movement of your body in order to shorten the
movement of the sword; by continuing you will reach the adversary's body before he can change his line. If when you reach
your position he changes his hand to _tierce_ or _quarte,_ then you
should parry on the inside and follow on.

Again if the adversary were in a low _tierce_ or _quarte_ directed downwards, you should place your point in the line of
his hilt, but above it, towards his hand, and,when you reach
your position,at once go on to the body, carrying your hilt to
the defence, for then he can do no harm with these low guards;
if he raises his point, it will meet your _forte_, as you are
moving. After reaching your position, you will pass so quickly

that he will have no time to defend. You must take care to
place your sword in position always with the hand in <u>quarte</u>
both on the outside and on the inside, above and below, and to
direct your point towards the adversary's hand and hilt, and
so far distant, that you always have time to disengage it or
change your line, before he touches it. The nearer to his body
you can bring your point with these precautions the better you
will succeed. Therefore to approach in the proper manner you
must carry your point forward without pause, and in such a manner that you can abandon your first plan and adopt another according to circumstances.

With this method you can make a feint of putting your
sword in one line and then put it in another. Therefore it is
well to remember that if your adversary, being in <u>tierce</u> or
<u>quarte</u> tries to hit in the time of your approaching, you should
always parry in the line in which you have put your sword;your
body too should be in that line, because if your body were in
one line and your sword in another you could be deceived, and
the method would fail; the sword must be accompanied by the
body and feet, and there must be no disunion. Also the parry
on the outside or inside must be made with the hand in quarte,
but the parry below and on the outside with the hand in <u>tierce</u>
so that the hand has little movement to make, and so that a
change of the hand is seldom needed.

Another method will follow, which is even safer and more
subtle, against which the adversary cannot use his left hand;

and sometimes against the other four where the sword is carried
in position in order to take the _time_ of the adversary's movement. But in the method of which we are speaking, the sixth
in order the sword is never so far forward that it can be grasped by his left hand, as we shall explain is its place.

<u>Pl. 142</u>.

With this <u>quarte</u> the sword is seen placed on the outside
with the point directed towards the adversary's hilt, whose
sword is in <u>tierce</u>. You have advanced from a distance with
short steps, carrying your sword in such a way that, on arriving
within distance, it is in the position shown. If your adversary
thinking he has the control over your sword, thrusts with a
change to <u>seconde</u> in order to meet your point with his <u>forte</u>,
your intention is to disengage the point with a slight movement;
and therefore you have placed it in this position. If your adversary makes no move, you will go on to hit as soon as you have
reached your position since your point will be very near his
body, and although it appears that he can control it, it is
nevertheless free; if he tries to engage, his <u>faible</u> will meet
your <u>forte</u>. All this is due to the continued movement without
a pause. If the adversary thrusts in order to hit the part uncovered over your sword, since your point is against his <u>forte</u>,
by simply raising it a little and extending the arm, you will
hit in <u>quarte</u> over his sword in the chest.

274

<u>Pl. 143</u>.

This hit in <u>quarte</u> has followed in this manner. Your adversary being in <u>tierce</u> and seeing you advancing to put your point against his hilt on the outside and that in advancing you have exposed your chest above the sword, has thrust with his hand in the straight line in order to hit in the <u>time</u> of your advance, and to cover himself also. You were advancing with the body and feet in union, holding your sword steady with the point near his hilt; you have disengaged your sword which was already low, with a slight movement, and by continuing with the left foot have met the adversary in his advance with the same <u>quarte</u>, carrying your hilt, already advanced, to his <u>faible</u>;in this way you will follow on to his body. If he had not moved, you would still have hit in <u>quarte</u> over his sword by moving the point a little and carrying it on with the <u>forte</u> against his hilt. He could have defended himself only by breaking distance, and in that case you would have dropped your point below in <u>seconde</u>, before he had raised his sword. If he had tried to save himself by disengaging in <u>quarte</u>, you would have continued in <u>quarte</u> without making any movement of defence, but with your hand in the parry of <u>quarte</u>, and would have hit at the same time.

275

Pl. 144.

With the next <u>quarte</u> the point is seen against the hilt of a <u>tierce</u> on the inside; the whole chest is exposed to the adversary. You have advanced from a distance with short steps, as is required with this method, and as is best with any other method, and whilst approaching have gradually brought your sword to this position, in order to provoke the adversary to attempt to hit or engage your sword. If he does not move , you will thrust the point at his body, after reaching his hilt, carrying your <u>forte</u> to his <u>faible</u>, and hit in <u>quarte</u> or in <u>tierce</u> according to circumstances. In any case you must continue,whether he offers a <u>time</u> or not, advances or retires. From this advantage will follow the hit shown in the next plate.

276

<u>Pl.145.</u>

This hit in <u>quarte</u> against an opponent attempting a hit also in <u>quarte</u> has arisen in this way; whilst you were approaching and carrying your point in the line of his hilt, the adversary seeing your point in such a strong position, has turned his hand from <u>tierce</u> to <u>quarte</u> in order to cover himself in the upper lines to engage your <u>faible</u> and hit. You were holding your sword steady and seeing his movement you have followed on, partly disengaging your point in <u>quarte</u> but without making any movement of defence, and have thrust in the angle formed by his hand in <u>quarte</u>; by running along his <u>faible</u> on the outside you have made the hit in the chest. You would have hit equally on the inside if he had not moved, or again on the inside with the same guard of he had tried to parry with his <u>tierce</u>. All this is due to your being in motion, which, as we have said elsewhere, leads to quickness, and causes your adversary to move without your moving your sword.

277

Pl. 146.

With this next _quarte_ the sword is seen to be placed on the outside with the point directed towards the adversary's hilt, who is in a guard of _seconde_. With this guard your chest is exposed to the adversary with the intention and design of enticing him to hit in the line uncovered, so that you may parry and hit in the same _time_. If he does not move, you will throw your point over his sword, and keeping the hilt steady hit in _quarte_, as shown in the next plate.

Pl. 147.

From the position of the two combatants in the last plate has followed the hit now illustrated. You have arrived within distance in _quarte_ with your point against his hilt; the adversary has not moved; therefore you have at once carried your point over his hilt; keeping the hand steady in _quarte_ you have brought your left foot in front, advancing the right side, and extended your arm; thus you have made a hit in _quarte_ in the chest, and by the natural strength of your sword in this line, you have pushed away his sword and prevented his parrying. You would also have hit in _quarte_, if he had disengaged to hit in _seconde_ on the inside, and you would have protected yourself by the some _quarte._

278

Pl. 148.

This next hit in _quarte_ has arisen from the some guard.When
you came within distance, the adversary made no movement and you
raised your point and carried it over his hilt in the manner described in the last discourse. He has disengaged in _seconde_ in
order to hit on the inside, imagining that you would have to
change your line to parry, and that by the strength of his sword
he would penetrate your body in the angle naturally formed by
his _seconde_, before you could parry. He has been deceived because, having already directed your point at his body, you have
gone on, merely carrying your sword arm towards the inside, and
extending the arm as in the last plate. You have met his sword
with your hilt, before he had finished his disengage, and therefore his sword is seen to be kept low; also your arm is foreshortened as it has been carried inwards for the defence. Similarly, when you first came within distance, the adversary might
have attempted to hit above in the part seen to be uncovered;
by disengaging _quarte_ you would have hit on the inside and parried in such a way that his sword would have fallen, since he
would have tried to force your sword, expecting you to parry,
but would not have found it because of your disengage, and
therefore his sword would have fallen, as we said. Further you
could have parried and hit above, as shown in the last plate,
even if the adversary had tried to resist , and this because
of the advantage of the line, since he would be resisting with
his weakest part against your strongest part.

279

Pl. 149.

This _tierce_ with the point against the adversary's hilt, the adversary being in a low _tierce_, is formed with the following design. You have seen that he was in a low guard with his upper parts exposed, although distant; your plan is to place your sword above his blade with the point directed towards his hilt in order to induce him to raise his sword on one side or the other, and to take the _time_ of his movement. If he does not move, you can go on to hit in that _tierce_, directing your point towards his throat close to his right arm and carrying your hilt near his sword in order to keep yourself defended. For in this position you cannot hit lower without moving your point from the correct position and endangering yourself.

280

<u>Pl. 150.</u>

From the advantage of the low guard of <u>tierce</u> of the last plate, with the point against the adversary's hilt, has follow-ed this hit in <u>tierce</u>. When you came within distance, the adversary did not move, so that you have followed on with the body and hit. You would equally have hit, if he had tried to it in any line, because you would have continued and been still defended, and similarly you would have arrived before he had finished his disengage, or your point would have reached the height at which your hand was originally, so that he could not have parried except by stepping back. As to his body he could not have withdrawn that further, and if he had tried to retreat and parry, that would have given you an opportunity to change your line; if he had parried on the inside, you would have yielded in <u>seconde</u>; if on the outside you would still have changed to <u>seconde</u>, but under his arm. All these devices will succeed if you follow on without stopping, remembering that to stop and then advance is very dangerous; rather than that it is better to retire and recover and then begin again.

281

THE SIXTH AND LAST METHOD OF ATTACKING THE ADVERSARY
WITHOUT A PAUSE, AND WITHOUT WAITING FOR A TIME BUT
FORCING HIM TO OFFER A TIME.

The preceeding principles and methods of attacking without
a pause, which we have described are practicable and likely to
succeed, some more than others; and of them the one which requires more skill is the best. The method of which we are now
about to treat is still more artful; in it the sword, feet and
body are used with greater subtlety than in the others. You begin by approaching from a distance with natural steps, as we in-
structed in the other cases; when you arrive within wide distance, the point of your sword should be against the adversary's
faible and in the stronger position. You should begin with the
arm advanced, and, as the body comes forward, the sword hand
must approach in order to bring the point to the required position of advantage, when within wide distance. In brief, although the body is moving, the sword and arm must remain steady
and the body approach in a particular manner. As to the feet
as you put one to the ground you must lift the other and bring
it up to the first, but keeping it in the air in order to put
it to the ground wherever needed, if the adversary moves. If
he does not move, you should put the foot down a little in
front of the other, and immediately lift the other to the same
extent and keep it in the air with the same intention; for if
the adversary takes the time of your lifting it, you have time
to adopt a plan, before your foot comes to the ground; if he

takes the _time_ of your putting it to the ground, the other foot must be in the air ready to move. In this way you are always on one foot only and can move quickly and with control, as you desire. You must take care always to carry the feet in the _line_ of the adversary's sword; if his sword is on the inside and possibly high, in bringing your body towards his arm you must raise your hand just enough to bring your point above his and keep it there; if in that _time_ he tries to hit in the lower lines, where he has been driven, it will be convenient for you to parry and hit below in the same _time_ on the outside, if your feet are in a straight line with his sword; in the execution of this movement your hand must be carried to _tierce_, your left side forward, while your right side is drawn back, for two reasons; in the first place, if the adversary disengages, he will not find your body; in the second, the further advanced the left shoulder is, the stronger the sword is and the more it can be shortened, so that you can advance further within close distance, and similarly your body will pass the point of danger.

With this method there is one rule which must be observed: you must keep your sword in the position where it engages the adversary's until you hit, nor must you hit, if your body has not passed his point; or not begun to pass; even if, at you advance, he should make some change, you must still contain yourself and not hit, unless you clearly know that your body can penetrate his point. Otherwise it would be better to seize the advantage in the other line, and without moving the arm or

hand, but merely by a movement of the body and a slight movement of the point, to continue on so that you could reach the adversary without entirely extending the arm. For with this method there is another rule also to be observed: you must carry your point to the adversary's body in union, without ever extending the arm or moving it. In this way you will be always ready with your body, feet and sword to adopt any change, and hit with safety and with great force, for it will be the body which hits, and not the arm. In this way either his sword will pass or break or he will be thrown to the ground; there will be no question of his being able to pass your point with his body or carry his body out of line, or of beating your sword or parrying with the left hand, with the sword alone.

With the sword and dagger this method may be used in some cases, though not in all. For if the adversary's point were in conjunction with the point of his dagger, you could not engage it without danger of losing your own sword. In this case it is convenient to adopt the other method of placing your point against the adversary's forte, where you can keep it free and hit with less subtlety than with the sword alone, since your sword is further from the adversary's sword and safer from his dagger, which cannot engage owing to the distance. You must take care in advancing to adopt a particular manner so that the point of your sword may not penetrate and may be kept always in the same position; as your body advances, so your dagger must

approach his sword in such a way that, when you are ready to hit the dagger will be so far forward that it can defend against his sword without further movement. With the sword and dagger also you can attack without stopping, but you cannot use one method only against all positions, as with the sword alone, but you must adopt now one, and now another according to need. In order that this may be better understood, we shall give plates of the positions, showing the manner first of placing the sword, and then of the hit, which may follow, as at present we show the positions with the sword alone.

All these methods of attack with resolution are based on the advantage of the feet, body and sword. But if the adversary does not keep his point steady but continually moves it round in a circle, it is not easy to engage it. In that case you could exclude his sword and prevent his moving it, though in truth this remedy involves the danger of your being disordered; therefore it is much better both with the sword alone and with the sword and dagger to advance holding your point in line with the adversary's hand and continue, since he will be forced to stop his movement and to try to drive your point out of line, otherwise you will go on and hit in the _time_ of his moving his point, and he will be unable to parry with his sword or dagger, if he has a dagger, since your point will be far distant from his dagger and far advanced towards his body; if he tries to carry his dagger to the defence of the other side, he will not be able to parry and will give you a chance to hit because of

his slowness due to the great distance. Therefore there is no position in which the adversary can place his body and weapons, to which with these methods there is no reply, since you have an advantage.

There are some men, speaking with rashness rather than with knowledge, of this art, who have presumed to say that there are some strokes, to which there is no reply,and which cannot be parried. But we are persuaded by sound reasons that every stroke has its reply, except the stroke made in the exact time and at the exact distance; to such a stroke there is no reply and it cannot be parried; whereas the stroke which is deceived in its time or its distance, has its reply and can easily be parried; so that you can defend yourself against all strokes of the one kind and none of the other, and he who thinks differently is deceived. So are they deceived who think that the same stroke can be used against every opponent. But we say that you can attack all opponents, but must proceed in different ways according to the opportunities offered by the adversary. Let this suffice for the methods of attacking with resolution and without a pause; you must understand how to advance or check yourself, move swiftly or slowly or retire, and to do everything of your own accord and not under the compulsion of the adversary, for that would be a sign that his replies were stronger and that you were trying to save yourself from danger. When you act of your own accord or for some purpose of deceiving, you can return and advance at will. In this consists true judgment and

knowledge of arms, which gives you the assurance of proceeding according to the capacity of your opponent and according to the position in which he is. We have still to give the description of each plate beginning with the advantage acquired and the distance, and continuing with the hits which follow from those advantages and distances.

287

Pl. 151.

With this method you have begun to advance against the adversary and have acquired the advantage shown in the plate.
You have carried your sword in such a manner that when within
distance you have acquired the advantage on the inside with a
guard of quarte against a tierce . This position has followed
from the fact that the adversary was more exposed on that side
To strengthen your sword you have turned your body and enlarged
the angle naturally formed by the hand in quarte; your front is
so turned that you are secure on the inside and there is little
exposed on the outside. For this reason you have turned the
body, so that you can defend yourself with little movement. This
position of the body strengthens the sword on both sides much
more than if you had your right side forward. You can follow
on with the left foot in order to approach without advancing
                                              in
your sword more than is now shown; this will be seen/the next
plate.

288

Pl. 152.

From the first advantage seen in the last plate has arisen the advantage now shown. The point of your sword is in the same position as before; you have not run along the adversary's sword, but checked your arm and advanced with the foot and body only; in doing this you have carried back the right side and advanced the left, so that you have in the end brought your head further forward than your hand.   Your intention is to lift the other foot, carry it forward and on bringing it to the ground to thrust, bringing your hilt somewhat further than the position where your point is at present; thus you will ran along his blade and penetrate with your body right to his body in such a way that he cannot prevent you.

289

<u>Pl. 153.</u>

From the two positions of advantage described in the two
preceeding discourses has followed this hit in <u>quarte</u>. Although
the adversary has tried to draw back and form a guard of <u>quarte</u>,
he has been unable to complete the turn of his hand, before you
have hit, because he has allowed you to approach too far, before
he moved.   After you had acquired the second advantage,   there
was no time for him to parry on that side;   but if he had moved
when you had reached your first position, and engaged his sword
he could have disengaged, not in order to hit, since he could
have effected nothing, but to engage your point on the other side;
he would have freed himself from the first danger and put you
under the necessity of using great swiftness of hand in order
to direct your point and exclude his on the outside;   before he
had finished his disengage or counter-disengage, and in order
that you might approach with your body without advancing your
sword more than in the first position. Your proceeding would
have been rendered more difficult;   but if you had worked in
the correct manner, he still could not have saved himself in
the end, because of the force and strength of this method of
attacking, for the nearer you approach the adversary, the safer
you become by reason of the turns you may make and the union
of body, sword and feet.

290

Pl. 154.

In this plate you are seen to have carried your sword in such a way, that on reaching the adversary's sword you have acquired the advantage on the outside with a guard of <u>quarte</u> against a <u>tierce</u>. You have taken up this position for two reasons, in the first place in order to be stronger in the line where your adversary's sword is, in the second in order to protect your body in the part exposed by the angle made by the guard of <u>quarte</u> near the hand. Your body is turned to the front exposing the whole of the chest; you are safe on the outside and with your hilt in <u>quarte</u> you are almost entirely covered on the inside, so that you can defend with little movement in any line. You have lifted your rear foot in order to carry it forward without advancing your sword more than it is at present, as will be seen in the next plate.

291

<u>Pl. 155.</u>

From the first advantage shown in the last plate has followed this further advantage. Having reached the adversary's
point and acquired the advantage you have continued with the
left foot, so that your sword has advanced no further than it
was. You have carried forward the left shoulder, keeping the
right shoulder in its original position, and thus have secured
yourself and deprived your adversary of the chance of hitting
below in any way, while on the inside you are covered by keeping your hand steady in its present position and by bringing
the right foot in front, if necessary; thus you have nothing
to fear in those lines, and on the outside you are similarly defended, so that in this position you can go on to hit in the
part seen to be exposed from the <u>faible</u> of his sword to his body
as we shall show in the next plate.

292

Pl. 156.

From the two positions of advantage just described has followed this hit in <u>quarte</u> against an opponent who has tried to
parry in <u>tierce</u>. On reaching the second position, as the adversary did not move, you have continued with the body without advancing the sword more than shown in the plate, still keeping
your hand in the guard of <u>quarte</u>. All this is done with great
skill, because on reaching the second position with the arm withdrawn if you had extended it to hit, you would have given the adversary a <u>time</u> to hit below in <u>quarte,</u> and to turn his body, letting your sword pass in vain, or to parry without disengaging.
In extending the arm the sword in fact is weakened and may be
easily thrust aside by the adversary, but when it is accompanied
by the body the adversary is not strong enough to drive it away.
For this reason then you have maintained your guard of <u>quarte</u>,
and also in order to parry more easily, if he should try to hit
below. You have lowered the body in the straight line in order
to render the defence easier both below and on the inside, so
that wherever he should hit, you would defend with little movement of the hand and body; moreover you would be so far advanced
that his sword would pass, and you would be out of danger. On
the other hand if you had bent the body outwards, you would have
been more exposed on the inside and would not have advanced so
far with the body, so that your adversary could more readily
have recovered his sword, whilst you would have been less united;
for all these reasons the movement would have been weaker.

293

We might have included the results which would have followed against guards of prime and quarte, and also guards at an
angle, or withdrawn; but we have omitted them for the sake of
brevity, and because whoever can advance with safety against a
guard in the straight line, can more easily attack those at an
angle or withdrawn.Therefore we shall not treat of them, since
they may be readily met with the methods we have described;for
the nearer you can approach the adversary before being impeded
or checked by his sword, the safer you are and the quicker you
will attain your end; the adversary has fewer resources when
you are close; when the danger is greater, he cannot make many
changes. As to rushes which may be made by guards at an angle
or withdrawn, we omit those also, because they will give no
trouble; for if you know how to attack according to our methods,
you will always be covered in the straight line from the adversary's point to your body. As to the changes of line made by an
opponent using a guard at an angle, they are always slower than
with a straight guard; therefore in these six methods we have
described the opponent as on guard in the straight line. There
are some who claim that a straight guard cannot be defeated, especially if the body is held sideways, whereas we have here
shown in how many ways such a guard may be deceived.

We have still to add, that with the last method it is better
to use a shorter sword, which is easier to control, less likely
to be impeded and has less faible; if the adversary's sword is
longer, there is all the greater advantage in attacking with

resolution. If you understand these methods, you can attack any imaginable guard; as the number of guards is almost infinite we have been content to include the principal ones, from which you may easily understand how to proceed against any other. Here we shall end the discussion of the principles of the sword alone and shall proceed to treat of the sword and dagger.

295

BOOK 2. PART 2.

On attacking with Resolution with Sword and Dagger.

Having fully discussed the method of attacking the adversary without waiting in presence with the sword alone, we shall now treat of the principles to be observed with the sword and dagger. Although with these weapons you make use of the advantage of the continued movement of the feet, which gives greater quickness of action, still you most consider that you have two weapons to control, and that your adversary has two weapons, against which you must defend yourself. To proceed in the proper manner requires great judgment in understanding the advantages and the dangers. In attacking the adversary where he is uncovered, in order to make him move, you are in great danger of losing your sword, that is to say the danger that his dagger, if not his sword, will engage your sword, and not only impede your plan, but put you in peril, Therefore you must be careful not to go so near either of his weapons, that your sword may be deprived of its freedom; although it is true that the more you can advance your sword and still keep it free, the more successful your method will be, even though the danger is greater.

With these weapons you must not only contrive that your dagger defends with little movement, when the adversary's sword attacks, but also that your own sword is in such a position, that it may hit in the exact _time_ and defend the line nearest to the line in which the adversary is attacking, so that if

he has made a feint of hitting in that line in order to hit in
the other near line, which would be uncovered by the movement of
defence, he will find the path closed and defended; it is not
difficult for the two weapons to defend the two lines, the one
defending the line in which the adversary attacks, and the other
the line in which he may attack, and without impeding your power to strike in the same _time_; if other methods are adopted
you may be deceived. With these weapons also there are more
paths in which the adversary may attack, and in which you may
attack him, but there are fewer methods of proceeding against
his movements because of the impediment of his dagger. Still
there are four ways or methods by which you may attack without
waiting for a time or anything else, but may go on resolutely
without a pause. These involve three guards, which are illustrated in the plates, so that you may know that with them you
may attack or wait, as you please. Of these we shall now treat
more in detail; we shall begin with a low guard in _seconde_, formed with the sword across the body. Then we shall speak of the
others in order as we have done hitherto.

297

THE FIRST METHOD OF ATTACKING THE ADVERSARY WITHOUT A PAUSE WITH THE SWORD & DAGGER.

In this first method of attacking with resolution a guard of _seconde_ is used, which has been already illustrated (pl.56) with the sword and dagger. You begin with natural steps moving in a circle towards the left,as the principles of this guard require, so that you can withdraw or approach without changing the position of the body or the nature of the steps and can contrive to entice the adversary to advance. As we explained elsewhere in speaking of the nature of this guard and of the manner of holding oneself in forming it and working with it,the circular movement is intended to protect the part which is uncovered over the sword; by keeping your dagger outside the adversary's sword on arriving within distance and holding your sword close to your dagger, when your dagger has penetrated the point of the adversary's sword, you prevent his disengaging below. Since your sword has closed the path between the weapons and your dagger is held upright, he can disengage over the point of your dagger only by a very slow movement; nor can he make a feint and hit, which would cause disorder and a separation of your weapons, and perhaps lead to a hit. Your body is held low to prevent his being able to hit below. This guard is based on such a position that the adversary can hit only on the inside of the dagger over the sword,and consequently you must protect that part as you advance,taking care that in the meantime you do not uncover other parts so much that you cannot parry; having formed this guard you must make certain that the adver-

sary's sword cannot hit or harass you except in this line over your sword. You must protect this line, so that when you have reached his sword your body is entirely out of line of his point. In no case with this guard must you separate your sword and dagger you must take care that your dagger on reaching his sword is close to its blade, so that you may run along the blade without beating it in order to avoid the danger which might arise, if the adversary raised his sword from that position and placed it in another, or withdrew it and hit after your dagger had fallen, or disordered your dagger by a feint and then hit in the time of your trying to defend; moreover the adversary's sword is at once free after being beaten, so that you would lose the control of it. But if you merely ran along the blade, you can then follow it wherever it goes, secure against any movement the adversary may make; your security is all the greater because he can never engage your sword, which is crossed out of line. You should make no further advance towards the adversary, after bringing your dagger hand to this position, so that the _faible_ of your sword will be at a distance, out of reach of his dagger; it is true that it is within the reach of his sword, but only with the danger to the adversary of having his sword excluded and being hit, before he can free it; hence your security, since your sword cannot be engaged. For the rest you must know how to apply the following rules, when within distance; if you have to hit in _seconde_ you must leave your dagger against the adversary's sword and hit when the opportunity occurs; if you have

to hit in _quarte_ by a turn of the hand, you must thrust with the blade of your sword still close to the dagger, so that the path between the weapons will be closed; after completing the extension the sword hand should be still close to the dagger hand, whilst the body should in no case be raised, but lowered still more at the moment of hitting. If you observe these rules you will defeat any adversary.

-300-

<u>Pl. 157.</u>

Here we give a plate illustrating the guard is <u>seconde</u> of which we have spoken, in order that it may be better understood. From this you will understand the method of advancing and carrying the weapons and in what position the sword and dagger should be on arriving within distance. You must always contrive to have the dagger on the outside of the adversary's sword, as in the plate. If you cannot because he has carried his sword so far away as to be out of line, you should continue in that direction until your dagger is near the blade of his sword, without changing your guard, and then immediately attack and hit in the part seen to be exposed, still keeping the body low. If it is necessary to hit between the weapons, you should hit in <u>quarte</u> and here we warn you never to divide your weapons either when turning the hand or making any other movement, especially when his sword is on the outside; for if you turn to <u>quarte</u> and separate the hands, his sword could thrust below between the hands. If the adversary is on the inside, as in this plate,you can hit in <u>seconde</u> between the weapons, if there is an opening, or below or over the sword according to circumstances, leaving the dagger to defend, as we have said, and as will be seen in the next plate showing the hit.

301

<u>Pl.158</u>.

From the preceeding guard of <u>seconde</u> with the dagger on the
adversary's sword has followed this hit over the dagger.   After
engaging his sword you pushed on against the adversary, who was
in tierce, as shown, and disengaged in <u>seconde</u> over his dagger,
running along the blade of his sword with the blade of your dagger.   Seeing you disengaging and advancing he has carried back
his left foot and thrust with his sword in order to meet you; by
turning his body he thought he would be able to parry with his
dagger, but he has failed, because his sword was already engaged
by your dagger and you have protected a great part of the body,
which was exposed before over your sword in <u>seconde</u>;   you have
followed his blade, completely defending yourself, and hit without his being able to save himself.   This is due to your continued union when disengaging, so that when the disengage was complete, the hit was already made, and his dagger could not parry.
Similarly it may be that the adversary seeing you attacking and
uncovered above the sword, has tried to hit in that line. Having
already engaged his sword and knowing well that you could not be
hit elsewhere, you have pushed your dagger forward along his
blade, advanced and made the hit in <u>seconde</u> on completing your
disengage.

302

Pl 159.

From the same guard of _seconde_ against a _tierce_ has followed this hit in _quarte_. When your dagger reached the point of
the adversary's sword, he tried to free it by disengaging below
the sword on the outside; but you, who were in _seconde_, turning
your hand to _quarte_ and bringing the _forte_ of your dagger near
his point, which had gone below, have pushed on and directed
the point of your sword, which was crossed out of line,against
his body in such a way, that when in presence the point had already reached his body; he has tried to parry with his dagger
and change to _seconde_, but has found his sword excluded,while
your sword was so far advanced, that his dagger instead of meeting the _faible_ has met the _forte_ and failed to thrust it away.
You should remember that when you have begun to engage his point
with your dagger and he moves his point, that is always the
time to attack, still preserving your union, so that if he parries before you have reached, you can follow on and change your
line.

303

THE SECOND METHOD OF ATTACKING WITHOUT A PAUSE WITH THE SWORD & DAGGER.

With this method also a guard of _seconde_ is used, which has also been illustrated (Pl.57.) it is formed with the feet in line and the points of the feet facing the adversary and wide apart, the body bent forwards and the shoulders in line, so that the whole of the chest faces the adversary; the arms and weapons are held high and curved inwards, so that the point of the dagger meets the sword near the _forte_ closing the path between the weapons against thrusts and cuts, with the point of the sword directed to the left, so as to cover the head entirely and defend it from any cut without any further parry. Thus the adversary can attack only below between the weapons towards the face, which may be easily defended by both weapons, which are advanced. The lower parts also are defended and safe by reason of the distance, where the adversary cannot reach them except by bringing his head close to your hands.

When you have formed this guard you should advance with short steps in order to keep the same distance between the feet. When you are so far advanced that your hands have penetrated the point of the adversary's sword, you should then carry one of your feet into line with the other on the side on which you intend to attack, which will bring your body out of line a distance equal to half the space between your feet when on guard, and you will be so far advanced, that your adversary can no longer bring his point into line. If you have carried the left

foot into line with the right, you can hit in <u>quarte,</u> holding
the point of your dagger turned downwards in order to exclude
his sword, so that it cannot attack in the line of your body.
If you have carried the right foot to the line of the left, you
will have greater advantage in hitting, since in carrying the
body away you can leave the dagger to defend the side nearest
the adversary and direct your sword, which, as we have said,is
high and held across the body, against the adversary; your
sword will then be so far advanced that its forte will penetrate
before his weapons can parry, and the point on coming into line
will hit. If the adversary should attempt a rush, you should
either parry by leaving your dagger to defend without however
beating his sword, or carry your foot to the right and thrust
the <u>forte</u> of your sword forward in order to exclude his sword
and continue with the point to his body, meeting him at the very
moment of his rush; in this way your success will be greater
then when he remains steady on guard. If he remains on guard,
it will be better to move to the outside with the right foot,
when by the advantage of being already on the move, you will arrive so quickly that he cannot save himself. With this method
you are certain that the adversary can never engage your sword,
which is a very great advantage. If you observe these rules,
you can attack any imaginable position on guard.

305

Pl. 160.

This is the guard of _seconde_ of which we have spoken, with the chest facing the adversary's sword. When within such distance that your dagger has penetrated his point, you advance the right foot and ran along his blade with your dagger, passing out of line towards his right side in such a way, that his sword which was directed against your chest, is now out of line of your body as far as half the space between your feet, before you moved. Therefore it is sufficient to hold your dagger against his blade, without thrusting it away, attacking over his dagger for you know that your sword is superior and can make a hit in _prime_, as will be seen in the next plate.

306

Pl.161.

From the guard of _seconde_ described in the last discourse, with which you have advanced so far that your dagger has penetrated the point of the adversary's sword, you can understand the effect and cause of the hit now shown. When within distance you have carried the right foot to the line of the left beyond the adversary's right side, and thus brought yourself out of line and reached his body with a guard of prime over his dagger;this guard has followed from the position of the _seconde_ with the sword crossed and out of line; in directing the point you have not lowered your hand but raised it, as seen, preventing his dagger from parrying. Although the adversary has drawn back and tried to turn his body and has begun to extend his sword, your stroke has arrived before he has finished the extension; even if he had withdrawn still further, he would still have done no good, but would have been hit in the same place. If he had contrived to defend the first blow, he would still have been hit below with a second.

307

Pl. 162.

When you had reached so far with your guard of _seconde_ that your dagger had begun to penetrate his point, the adversary tried to disengage in order to free his point. Seeing his intention you left your dagger in its position, turned your hand to _quarte_ and hit between the weapons, arriving with such speed that he has had no time to parry. Your sword was outside his right side and so far advanced, when he disengaged, that it reached his body at the moment of his disengaging and directing his point, so that he had not been able to thrust your sword away with his dagger. If he had withdrawn in order to have room to parry, leaving your sword hand in the same position you would have dropped the point far enough to hit under his dagger hand at the moment of his expecting to parry, so that he could not have defended. Also if his sword had been lower, keeping the dagger arm at the same height you would have turned the point downwards and kept his feint out of presence.

308

Pl. 163.

When you came within distance with this guard of _seconde,_ with the feet apart and in line, the adversary raised his dagger, covering himself above, advanced his left foot and turned his hand to _quarte_ in order to hit in the part which he saw exposed between your weapons. Therefore you have carried the right foot to the line of the left, thus bringing yourself out of presence; leaving your dagger against his sword in order to exclude it, you have directed the guard of _seconde_ below his dagger arm, and thus hit at the moment of his putting his foot to the ground and will follow on to the body without stopping.

309

THE THIRD METHOD OF ATTACKING WITHOUT A
PAUSE WITH THE SWORD AND DAGGER.

In the third method, which we are now about to explain a guard of <u>tierce</u>, illustrated in plate 65 with the sword and dagger, is used. With that guard with the feet together, the body bent and the weapons divided and held high you await the adversary because of certain advantages of the guard, as we explained in full.You now use this guard at the beginning only while approaching the adversary from a distance, for when you are within distance the position of your body and weapons will be different. With this guard then you begin by carrying the left foot outwards, immediately lifting the right foot and carrying it also in the same direction; you bring your sword down from its position on guard and unite it with the dagger, carrying your left shoulder back, so that when within distance sword and dagger are in contact with the hand in a guard of <u>quarte</u>, so that the adversary cannot enter between your weapons; your sword is held extended in line towards his chest or face, and your left shoulder so far back that your whole body is behind line of your sword; your point must be maintained opposite the part exposed by the adversary and your sword so far from his weapons that you are sure of freeing it before he can engage it, but at the same time the point should be as near his body as possible; you should leave the sword in that position without pushing it further forward, and follow on with the feet, bending the body and beginning

to turn the hand towards tierce; as you turn the hand the dagger is extended and the point of your sword lowered in proportion, so that it can be disengaged; while doing this you should bring the left shoulder forward, without letting the sword hand drop and continuing to turn towards a guard of seconde. When you have arrived from quarte to tierce and are beginning to turn toward seconde, the point of your sword should begin to pass the adversary's sword and dagger, but without having been pushed forward; your dagger should then have reached the blade of his sword, and when your hand has reached seconde, the movement of disengaging should be complete and your point should hit above the dagger, unless his dagger were too high and he were covered in that line; in that case you should remain below his dagger arm and hit in seconde, as also if the adversary's sword were held back and his dagger advanced and the points united; if the points were divided and his sword held back,you would arrive, on dropping from the high tierce into line, close to his dagger on the upper side with the hand in quarte and the point of your sword would have penetrated as far as the fourth part of the blade, or little less; then you should join the dagger and sword hand, so that if the adversary tried to parry with his dagger and hit in quarte, while your sword was approaching your dagger could defend your right side, and you could turn your body and hand at the same time and hit in seconde under his left arm. If when you carry your sword close to his dagger,

the adversary does not move, and your point has penetrated to
the fourth part of his blade, you should then turn your hand
from _quarte_ to _seconde_, thrust your dagger towards his sword,
with a slight turn of the body, run along his blade and go on,
hitting over his dagger in the line into which your sword had
dropped; he would be unable to parry, because in turning the
hand to _seconde_ the point of your sword would incline so far
inwards and would be so far advanced that it would be nearer
the spot where it hits than the path into which the adversary
is trying to drive it; moreover it would be so strong that it
could resist the dagger without fear of being thrust aside. If,
while you are bringing your sword into that line, the adversary
should raise his dagger to cover himself, keeping his sword
steady, his sword would then be so far withdrawn that your dagger could not reach it; therefore you would have to move your
sword from that line, carrying it over the point of his dagger
into the line between his weapons, and hit in _quarte_, keeping
your dagger so close to your sword hand, that he could not enter
between your weapons; you must also maintain your hilt against
his sword and bend your head forward so that he will certainly
be able neither to parry nor hit, since his sword will be excluded. If the adversary's weapons are divided and his sword held
back you could also with your guard of _quarte_ pass the point of
his dagger between his weapons and go on, with your dagger and
sword hand in conjunction; from this position you would carry
the dagger forward towards his sword, turning your sword hand

and leaving it there until you are sure you can reach his body with that guard, or until he tried to beat your sword or engage it in order to cover himself between the weapons, which would be an opportune moment to hit over his dagger. If the adversary uncovers himself below, you should advance your dagger to his sword and hit in _tierce_. He will not be able to escape the three attacks, between the weapons, above and below.

These are the principles to be observed against guards which are withdrawn, whether the opponents sword is long or short, whether his weapons are united or divided, and whatever the position of his hand. But you must remember that if the adversary's sword is advanced, whether his guard is open or closed, on bringing your sword into line you must proceed to engage his advanced sword, protecting your body by bringing it somewhat out of the line in which it is exposed, so that if the adversary changes his line, you are already defended. In trying to engage his sword you must use the dagger also and never arrive within distance with your weapons divided, whatever his guard. All these movements must be carried out without checking your feet or sword; when once the sword has begun to fall, you must be either advancing it or turning it without pause. This is an excellent method, likely to succeed against any weapons.

313

Pl. 164.

From the high tierce with the weapons divided and the feet
together, as illustrated in Pl.65 with the sword and dagger,has
arisen the position of this quarte in the following manner: when
at a distance with this guard of tierce, with the weapons high
and the feet together, you have begun by carrying the left foot
away to the adversary's right side and bringing the right foot
forward, at the same time bringing your sword and dagger into conjunction so that when within distance your hands were in contact
and your sword had completed the change to quarte; you have also
gradually lowered the point so as to bring it into line and directed it in the line uncovered outside the sword; you have advanced the body, without allowing your point to penetrate any
further, but turning your hand towards tierce, carrying the right
side somewhat back and bringing your dagger near his sword; you
have also lowered the point of your sword, so that,if necessary,
you could disengage it, that is if the adversary had tried to
parry with his sword. You would continue the movement of your
hand from tierce to seconde and hit on the inside between his
weapons; if the adversary tried to carry his dagger down to
your sword, you would still turn the hand to seconde and thrust
past the point of his dagger; parrying with the sword and dagger together you would disengage your point below as it was already with little movement being carried downwards and would

hit under his dagger, with your body low and letting his sword pass outside your left arm, as will be seen in the next plate.

Pl. 165.

From the guard of quarte with the sword on the outside of the opponent's sword in tierce has followed this hit. You had begun with the high tierce, as previously described, and come within distance in the position of the last plate. The adversary tried to parry with his sword and dagger in conjunction; but as soon as you were within distance you began to turn your hand towards tierce, placed your dagger against his sword,which was advancing to parry and hit, letting your point fall low,so that the adversary failed to find it; you continued the movement of your hand to seconde, which brought your blade outside and below the adversary's left arm. If, when you directed your sword into the line uncovered, the adversary had tried to disengage and parry with his dagger, you would have thrust in quarte which would have prevented him from doing anything, except retiring when you came within distance, or changing his guard in order to make you change; if you had failed to seize the opportunity of the change, you would have given him a chance to hit, if he had followed, though it is true you might have halted and adopted another method.

315

Pl. 166.

This next <u>tierce</u>, which has excluded the opponent's sword in <u>tierce</u>, with his sword advanced and his hands in union, has followed from the high <u>tierce</u> previously described you have brought your sword down and your hands together while advancing and excluded his sword with both weapons, keeping your right side further back than your left, so that, if the adversary disengaged, you could advance your body without any further movement to defend it, and could pass on to the attack, hitting in <u>seconde</u> or <u>tierce</u>, as his disengage was high or low. If he did not disengage you could pass on to hit, as will be seen in the next plate.

316

<u>Pl. 167</u>.

From the preceeding <u>tierce</u>, which had excluded the opponent's sword has followed this hit. The adversary was in an advanced <u>tierce</u> with the hilts of his two weapons together; when he made no move on your advancing, you left your dagger against his sword on the inside, turned your hand to <u>seconde</u>, carried your point over his dagger and hit in the chest, without his being able to parry. Although he has drawn back his feet and body, he has been unable to free his sword, because you were too far advanced and your sword had penetrated already to his body, when he found it.

Pl. 168.

317

From the high tierce is derived this quarte also, which has
come into presence over the dagger of an opponent in tierce on
the left foot, with his right side drawn back in order to avoid
the hit; which you might make by disengaging, and with his dagger raised to parry. In this case from this guard of quarte
you would turn your hand to seconde, and, if the adversary did
not move his dagger, you would hit in the upper line, since
your sword would have already penetrated far forward, and the
angle made by the turn of your hand would carry it with force
against the inside of his dagger so that the dagger could not
thrust it away. If he raised his dagger to parry without disengaging his sword, which is on the outside, by dropping your
point below and keeping your hands in conjunction you would hit
with this quarte in the same time below, as will be seen in the
next plate.

318

Pl. 169.

From the advantage, which the _quarte_ derived from the high _tierce_ had acquired over the adversary's dagger on the outside, has followed this hit. When you brought your sword above his dagger, his plan was to engage your sword; he raised his dagger to parry and turned his hand to _quarte_ in order to hit by disengaging and passing; you were already on the move and have seized the advantage, while he was raising his dagger, and have hit at the moment when he expected to engage your sword above; he has been prevented from either disengaging or passing, and even if he had disengaged, he would still have been hit.

Pl. 170.

319

This _quarte_, with the sword between the weapons of an opponent in _tierce_ on the left foot, is also derived from the high _tierce_. If the adversary makes no move you would advance your body close to the hands and at once hit under his left arm in _seconde_, while your body would pass outside his sword, and you would leave your dagger against his sword; if he disengaged on the outside of your sword, you would still hit in _seconde_ below; but if he disengaged on the side of the dagger, you would hit in quarte, dropping your point far enough to pass under his dagger hand; whilst hitting in _quarte_, you would press your hands still closer together in order to exclude his sword. If he tried to parry with his dagger and thrust in _tierce_ below, when he saw your sword attacking in the middle, you would then raise your sword over his dagger and hit, as will be seen.

320

<u>Pl. 171</u>.

From the preceeding <u>quarte</u> derived from the high <u>tierce</u> has followed this hit. When your sword came between his weapons, the adversary who was in <u>tierce</u> on the left foot, seeing your sword near his dagger, tried to exclude it with his dagger, and hit below in <u>tierce</u>. You, who were in <u>quarte</u>, as we said, turned your hand to <u>seconde</u>, and at the same time carrying your point over his dagger hit the adversary, as he lowered himself and advanced his foot. You could also have disengaged below his dagger hand and made the same hit over the dagger, but a little lower because of the angle made by the arm.

321

THE FOURTH AND LAST METHOD OF ATTACKING
WITHOUT A PAUSE WITH THE SWORD AND DAGGER.

This is the fourth method, in which we explain the manner of proceeding against those who never hold their sword still, but continually move the point in a circle and hold the dagger, now advanced, now withdrawn, at one moment close to the sword, at another separated from it. We must proceed in a manner different from the other three methods. If the adversary moves his sword in a circle, as long as his sword hand is at a distance from his dagger hand, you should begin to approach your points and advance as far as possible towards his sword hand, in order to force him to one of two course, either to stop moving his point and meet your advancing point, or to move his dagger, both opportune moments to hit in the part he uncovers. You should continue to hold the point of your dagger directed towards the adversary, so that it may be ready to parry on every occasion of his attempting to hit; when he must pass the line of your dagger. You must also keep your body low and in union with your weapons. If the adversary makes no move, you must advance so far that you can take the _time_ of the circular movement of his point, and hit, excluding his point at the same time, assured that, while your point is directed towards his hand, you will easily parry, whenever he tries to hit during your advance and will hit without being disordered. But the correct principles must be observed.

322

In case the adversary while moving his point in a circle has his dagger close to his sword hand, you must be careful not to advance your sword so far that it would be engaged. The true method is to keep the point of your sword directed towards the first part of the adversary's blade, that is towards the hilt, and to follow on by bringing up the body to the first position of the hands, with the hands now a little in advance of the body, bending the arms at the elbow, as you advance; for if you work from the shoulders only as you advance your body, your hands will go too low and you will be exposed. Your hand should be in quarte, with the points still directed towards the first part of his blade. As you come forward in union, you must lower the body, but never let the point of your sword penetrate so far that it is inside the adversary's dagger, that is, when his dagger is close to his sword hand. You should approach towards the blade or point of his dagger, and if he hits in the time of your advancing, you will parry more easily than if your points were directed towards his hands, since the fortes will be further from the adversary, so that the point of his sword cannot so easily penetrate them, and the fortes will be all the stronger, since they are gathered in close to the body. There is one point to be considered,that is that when his point moved in a circle, the first part of his blade also moved, and changes its position so much, that you cannot keep your points exactly against it. If you follow our rule and keep the points of your

sword and dagger directed towards that part with your dagger
hand somewhat divided from the sword hand, but with the point
of the dagger close to the blade of the sword, that wavering
of the adversary's sword will not matter, since it will never
be so great that he can hit in any line, disorder you, or obtain any advantage.

If the adversary holds his dagger so far forward as to cover the whole of the first part of his blade, and moreover close
to his sword in a guard of quarte, you cannot then approach the
hilt or the first part of his blade. If you tried to approach
the second part of his blade, you would not be safe, since that
part makes a large movement, when the point is moved, and the adversary would be too far away to be hit. In this case you should
hold the point of your sword against his dagger hand, and your
dagger not much advanced and directed towards the centre of his
blade; you should advance with your feet and body towards his
dagger side, with your hand in quarte and as you advance gather
your hand in towards your left side, still keeping the point in
the same line; when you have brought the hand as close to your
body as possible, you would then be in the required position and
could hit in the line uncovered nor would the adversary be able
to parry in any way. If your dagger is directed towards the
centre of his blade, it will easily parry if he tries to hit in
that time: if he hits before you reach your position, it will
be still better for your defence, since both your weapons will

be free and steady, so that when once within distance you could
take any _time_ offered by the adversary. In following this method
you must continue with the feet, keeping your points and hands
steady, until you find a _time_ or reach your position.

These rules may be followed against those opponents, who keep
their hands steady on guard and move the point in a circle, if
they change their hands from one guard to another, they would
offer an even greater time and could more easily he attacked.
But there are others who move the sword and dagger together, now
advancing the dagger, now raising it, now lowering it or
withdrawing it, continuing the circular motion with both weapons,
and even with the feet, with their weapons in conjunction and the
sword much in front of the dagger. Against these also you can
advance with your points directed towards their hands, but still
it would be better to keep the point of your sword and hand as
high as the adversary's dagger hand, when at the top of its
movement, for that hand has to guard against your two blades
together; your point should be advanced, but not so far as to
penetrate his dagger; you must advance resolutely, keeping your
dagger directed towards the centre of his blade; you will come
so far forward that your point will thrust in the line exposed,
when the adversary lowers his weapons, and will hit without his
dagger being able to parry, whilst you will be defended by
your dagger from the thrust in _quarte_ or _tierce_, which he might
make in that _time_. If on your advance the adversary makes some

movement in order to engage your sword with his sword or dagger,
or with both together, that also would be a suitable _time_ to hit.
The fact of your being on the move and having your weapons
steady is of great importance in carrying out their rules and
taking a _time_.   The purpose of these rules is simply to instruct
you how to reach a position where you may take a time, and, when
you are in that position that you may understand what may happen
and may be able to attack even if the adversary does not move.

Here we conclude the discourses on resolution, which we
promised. If we are not deceived they will be sufficient to
enable you to deduce other rules and methods; we have omitted
the numerous varieties of methods and merely considered the
foundations of the art, explaining its true principles, and
leaving room for some rare genius to add and discover other
methods. As far as possible we have avoided prolixity in order
not to weary the reader and we have refrained from using geometrical terms in order that the reader may understand more easily, although the principles of our art are based on the art of
geometry.

The plates of the fourth method of attacking with resolution with the sword and dagger will follow in order, and will
illustrate the effects and causes of the advantage acquired
and the hits made.

326

Pl. 172.

With the _tierce_ shown in this plate you have placed the point of your sword close to the blade of the adversary's dagger with the point of your dagger directed towards the centre of his blade, because he is moving his sword and dagger in a circle, successively raising, withdrawing and lowering both his weapons together. You intend to approach so far that you can take the _time_ of his two weapons falling, and therefore have placed your point near the blade of his dagger, which is now high; when he drops or withdraws both his weapons, your point will be free and his body exposed, so that you can easily make the hit shown in the next plate.

327

Pl. 173.

From the last _tierce_, with the point of your sword close to
the blade of the adversary's dagger, has followed this hit. When
you came within distance, the adversary raised his weapons and
then lowered them; as your point was close to the blade of his
dagger, you have thrust in the _time_ of his dagger falling, and
have hit changing your hand to _seconde_, as shown. He has failed
to parry with his dagger, because it is impossible, while you are
making one _time_, for him to make _two_, one in withdrawing his arm
and the other in returning it. Since this arm is not extended, it
is seen that his sword had not finished its advance before he was
hit. You have left your dagger in its original position in order
to defend in case of his sword hitting; therefore your arm, which
was first extended, is now withdrawn for, whilst your body was
advancing, if you had kept the arm extended, the point of your
dagger would have reached his hilt, and by thrusting he would
then certainly have hit, for two reasons, the first because your
dagger would have met the _forte_ of his sword and therefore could
not have resisted the impact, and secondly because by carrying
your dagger forward, you would have offered a _time_, all the more
because your body would not have been out of line, as it is now.
For these reasons you have maintained your position in order to
be ready for all eventualities, both for taking the _time_ of his
dagger falling, and for taking it when his dagger advanced, and
hitting below without his being able to parry. Also you could
carry your sword above the adversary's sword, when his weapons
fall, and prevent his raising it again except by a disengage.

328

If brief you have many chances of hitting and doing other things in order to disorder the adversary during his movements; but we mention only the more subtle and important.

Pl. 174.

The opponent is here shown in _tierce_ with his weapons divided moving the point of his sword in a circle, now carrying it away from and now bringing it near his dagger which is kept steady. You have brought your sword down from the high _tierce_ and directed the point towards the first part of his blade,without touching it, but with the intention of distracting the movement of his sword, and forcing him to do one of two things, either to stop his movement and meet your sword, as it advances, or to engage your sword with his dagger and hit at the same time; in both cases you would take the time and hit according to circumstances; if he had attacked your sword on the outside, by a slight movement of your point, which you have brought near his blade for this purpose, you would disengage and hit on the inside in _quarte_; if he attacked your sword with his sword and dagger together you would not disengage, but turn your hand to seconde, and parrying with your dagger hit below or above according to the height of his weapons. If he had tried to engage on the inside, you could have disengaged in _tierce_ on the outside over his sword, and while he was parrying with his dagger could have turned your hand to _seconde_ and hit over the dag-

ger, disengaging on either side. Or, while he was trying to engage your sword on the inside, you could have turned your hand to _seconde_, so that he would not have found your sword, and hit where the opening seemed best, but without stopping, for by stopping you lose your advantage.

Pl. 175.

From the _tierce_ with the point of your sword directed towards the first part of the adversary's blade when he was in _tierce_ and moving his point in a circle, has followed this hit. Seeing your point so far advanced, the adversary has tried to engage it with his dagger; but your point has penetrated so far and your hilt is so high, that by simply turning your hand and directing the point against his chest, with your hand still at the same height, you have hit at the moment when he expected to engage your sword. You have made this stroke at such a distance from his dagger, that it was impossible for it to return and parry; even if he had carried his dagger to that part, he would still have been hit in _quarte_ as will be shown in the text plate.

330

Pl. 176.

This is the quarte mentioned in the last discourse, hitting between the weapons of an opponent also in quarte. You have placed your point against the first part of the adversary's blade, when he divided his weapons in making that circle, of which we spoke in the general discourse; you have reached your position at that moment, and, seeing the opening and that he was making no further move, have hit by turning your hand from tierce to quarte, arriving before he could parry. If he had thrust forward in order to parry and hit in quarte, he would have done no good; the only effect would have been that you would have arrived sooner and the hit would have been stronger, whilst your dagger would have defended more easily, since it would have approached the point of his sword. It may have happened that you had reached the first part of the adversary's blade and found little uncovered; you have made a feint of attacking below the point of his dagger, still continuing your advance; the adversary has tried to parry by raising his dagger and turning his hand to quarte; in that time you have returned to the middle and hit in quarte, parrying as shown. If he had returned his dagger to that line in order to parry, he would have failed.

331

Pl. 177.

From the position of your point against the first part of
the adversary's blade has followed this hit. Seeing your point
so far advanced in the opening made by his circular movement of
his sword, he has tried to cover himself by uniting his dagger
and sword; taking that _time_ you have changed from _tierce_ to
seconde, disengaging over the point of his dagger; he has tried
to parry with his dagger, turning his hand to _quarte_, and advancing to meet you with his sword; he has not succeeded, because
the point of your sword had already reached his body at the moment of disengaging over his dagger, and because his _quarte_ was
easy to parry, since you had placed your dagger against his
sword, engaging it at the beginning of your movement, and running along his blade. Or it may be that you had placed your
point against the first part of his blade and thrust in the opening between his weapons; he has tried to parry and advanced in
order to hit below; you have gone on merely changing from _tierce_
to _seconde_ and met the adversary, who was also advancing.   He
has been unable to parry or hit, since your dagger had already
reached his _faible_, and, although he has tried to turn his head
to _quarte_ he has done no good.

332

Pl. 178.

From the preceeding <u>tierce</u> with the point of the sword against
the first part of the adversary's blade has followed this hit.
When within close distance you have taken the <u>time</u> offered by
the adversary in carrying the point of his sword away from his
dagger in its circular movement and have disengaged between his
weapons in <u>quarte</u>; he has been unable either to parry with his
dagger or to turn his hand to <u>seconde</u> because of the advance of
your sword, which had already hit when he tried to parry; for
this purpose he bent his body thinking to escape the imminent
danger, but when he turned his hand to <u>seconde</u>, your body had
already passed. You have hit with the dagger also at the same
time, while he was occupied in the effort of defending himself
from your sword, and because he was so impeded, that even if he
had tried to hit with his dagger, he could not have done so,
because his arm would have been imprisoned by your arm, which
had passed so far forward, that he could hardly have seen anything. This hit with the dagger has been introduced to show
that you can also hit with the dagger; if we have not spoken
of it before, although there has often been an opportunity, it
is because we have deemed it better to confine our attention to
the use of the sword. Moreover those who pass with resolution
have no need to hit with the dagger or to fear the adversary's
dagger, because when you pass and hit the sword penetrates entirely and removes all danger. Therefore you can pass without
fear of his dagger, assuming that no one is so foolish as to

let your sword pass through his body in order to hit you with
his dagger; even if an opponent did that, he would generally
be thrown to the ground before he could hit. Moreover since he
is forced to parry with his dagger, he cannot hit in time,
whilst on the other hand by advancing with resolution, when the
adversary's point is passed, you can leave it without hesita-
tion, and carry your dagger to his body.   Therefore it is clear
that he who passes can hit with the dagger better than he who
waits, whose lack of resolution is increased by seeing his opponent close upon him and his sword engaged, so that he can
parry with the dagger only; his dagger being engaged on one
task cannot perform the other.   Therefore he who passes has always the advantage, and if he does not hit with the sword, can
hit with the dagger, but if he hits with the sword, he will
not need the other. We might already have treated of this
manner of hitting, but our intention has been to consider the
point of the sword, which attacks from a greater distance,
takes and offers the times of hitting, and also is the first
to strike terror and attack. For these reasons we have desired to consider a subject, which is more subtle and profitable.
We have added this short discourse to show the error of those
who reject the pass from fear of being hit by the adversary's
dagger. We have also omitted for the sake of brevity the consideration of the broad-sword and many other kinds of weapons,
of which there would have been much to say. Moreover such

arms are not used among gentlemen nor in chance meetings, though
they are excellent when campaigning or   but such
matters are far from our subject, since we intend to treat only
of the arms of gentlemen and of cases which may arise in the
association of noblemen. Of these things we believe we have
treated at sufficient length; it remains only to throw light
on some extraordinary accidents, which may arise, although
rarely. For this purpose we shall add another short discourse
showing the method of defence on such occasions.

335

BOOK 2.   PART 3.

Treatise on Coming to Grips, Seizing the sword,
Throwing the Cloak, and Principles of the dagger.

Although our intention was not to treat of the following
Matters, because it seemed to us that our work could very well
stand without them, nevertheless owing to the persuasions of
many friends and to gratify them we have been induced to include
in our book this treatise on coming to grips, seizing the sword,
throwing the cloak and the principles of the dagger, that is
the principles of defence against the dagger with the bare hands.
The reasons which at first dissuaded us from treating of these
matters were the fact that the volume was sufficiently long
without them, and the fact that our purpose was to show how to
defend oneself and attack the adversary with the sword alone,or
the sword and dagger, or the sword and cloak, these being the
usual weapons among gentlemen and truly appropriate for noblemen; therefore we have said nothing of shields and bucklers
and other things, which it would take long to discuss. We have
always been of the opinion that one, who understands the rules
we have put forward, can use his sword in company with any
kind of weapon, whether in the hand or on the arm, for in all
cases the observation of the _time_ and distance is required.
Thus we proposed to treat only of the thrust and the cut, be-
lieving that, whoever can defend and attack in _time_ with these,
would never need to come to grips on the seizing of swords.

336

For similar reasons we have said nothing of defending against the dagger with the bare hands, since, when honourable men are driven by a point of honour to have recourse to a trial of arms they must do so on equal terms and with a correct test of valour, and should abhor a victory won by an unworthy and disgraceful advantage of weapons. Nevertheless we have accepted the advice of our friends, and since among men entirely honourable unexpected cases arise, and so suddenly that there is no time to resort to swords, it appears well to record how the dagger may be used against the dagger. Since that weapon is short, there is a danger of the adversary's seizing the dagger hand. Therefore, if possible, you should avoid parrying, and protect yourself by swift movements of the body and feet to one side or the other, hitting at the adversary's hands and arm; this will be a safer method and will keep him at a distance, so that he cannot seize your weapon or come to grips. For the rest if you have a knowledge of _time_, distance and passing, the rules already described will serve; therefore we say no more. But to satisfy one who can command us, we shall explain how to defend and attack, when assaulted by an opponent with a dagger, when you are unarmed.

337

Pl. 179.

The struggle shown in this plate may have arisen in the
following manner: you were in _tierce_ on the inside, and your
adversary also in _tierce_; having the advantage you have attach-
ed in _quarte_ close to his sword; he has tried to defend with
a _quarte_ and carried his point out of line; therefore you
have yielded your point, advanced the left foot and followed
with the right behind his right foot, bringing your hand above
his hand at the same time, whilst your pommel has reached his
chest as your foot came to the ground. Or it may be that the
adversary was on the outside of your sword; you have moved to
engage his sword, and taking that _time_ he has cut in _mandiritto_
at the head. Therefore you have brought forward the left foot,
parrying in order to hit in _quarte_; seeing his danger he has
attacked your sword in order to force it out of line; you
have yielded your point, brought your pommel over his sword on
the outside, passed and come to grips, as shown. Or it may be
that you , the assailant, had cut in _mandiritto_, and the adversary had made the simple parry of _quarte_, carrying his point
outwards; you, who had brought forward the left foot when
making your first hit, have passed in the time of his parrying
without finishing the cut, but bringing your hand above his
sword on the outside; that you have made the stroke shown so
that the adversary is on the point of falling to the ground.

338

Pl. 180.

This plate shown the wresting of the sword from the opponent's hand, accompanied by a thrust in _seconde_ in the chest; this may arise in two ways; in the first the adversary had cut in _mandiritto_ at the head; you parried in _seconde_, completely covering yourself, and immediately after parrying passed; putting your left hand reversed on the inside of his hilt and hand you have twisted his arm, turned it outwards, and wrested his sword from his hand by force, so that he has been unable to hold it. In the second method, you were in _tierce_ on the outside of the adversary's sword and have made a feint of hitting in the face, raising your hand to _quarte_ and carrying the point to that line; seeing his danger he has raised his sword to defend; you have yielded from _quarte_ to _seconde_, brought the left side as far forward as the right, lowering your head so far that the hilt and forte have entirely covered it, so that his point has passed behind; at the same time you have disengaged your sword in _seconde_, placed your hand reversed on his sword hand, and by bending his arm outwards forced him to relinquish his sword.

339

Pl. 181.

This plate again shows the wresting of the sword from the opponent's hand arising in this manner; you have offered the adversary a _time_ to hit on the inside in _quarte_; on his advancing you have placed your _forte_ over his point forcing it down, and at the same time brought your left foot forward, extending your left arm over your sword, placed your left hand under his hilt and lifted it upwards; by pressing down his point with your sword you have forced him to relinquish his sword. Or it may he you had made a feint of cutting in _mandiritto_ at the head; the adversary tried to parry and hit in _quarte_; abandoning the cut and letting your _forte_ fall on his point and pressing it downwards, you have seized his hilt with your left hand, as explained, and wrested the sword from his hand.

340

Pl. 182.

In this next struggle the opponent is hit in the chest. You had made a feint of hitting in <u>tierce</u> on the inside, he moved to parry, but you disengaged in <u>seconde</u> before he touched your sword; resting your left hand on your hilt in order to strengthen it, so that he could not thrust your sword aside,and bringing the left foot forward and behind his right foot, you have hit in the chest, putting your left hand reversed, which was on your hilt, on his throat, and forcing back his head in order to throw him to the ground. Or it may be that the adversary had moved to engage your sword on the outside in <u>tierce;</u> you disengaged in <u>quarte</u> and in the <u>time</u> of his parrying you yielded your point to his pressure, brought your pommel over his sword and turned your hand from <u>quarte</u> to <u>seconde</u>, so that you have passed over his sword on the outside, and at the same moment brought your left foot forward and made the hit at close quarters as shown.

341

Pl. 183.

Next follows a cast of the cloak. You had your cloak around you; when you had to draw your sword, you let the cloak fall from the right shoulder, leaving it on the left only; after drawing the sword from the scabbard you took hold of the edge which was hanging with two fingers of the left hand, and then with the left hand gathered it up close to the hood, as though you meant to throw it over your arm. Since you were so far from the adversary's weapons that more than one step was needed to bring you within distance, you have thrown the cloak over his hands, retaining hold of the edge. This has caused the cloak to fall over his sword so that owing to its weight he can raise neither the point nor his hand; thus by carrying the left foot forward followed by the right you have hit as shown; you have kept the edge in your hand, not only in order to extend it and cover the whole of his sword, but in order also that, if the cast missed its object, you could recover with a jerk over your own, and cast again either over his sword or in his face. If you had merely wanted to prevent his hitting or doing anything else, you would have thrown it free without retaining hold; but in that case it would be necessary, first to engage his sword.

342

Pl. 184.

This plate shows the casting of the cloak accompanied by a thrust in the face. You were in a guard of <u>tierce</u> on the left foot with the cloak around your arm; since the adversary's sword was also in <u>tierce</u> inside your cloak, you have rested the point of your sword against your cloak and carried it forward from the left hand with the help of a jerk; by bringing the right foot forward and accompanying the cloak with your point to his face, you have hit in the same movement. If he had tried to raise his sword and draw back to save himself, he would have effected nothing owing to the unexpected nature and novelty of the stroke, never thinking that you would throw your cloak, or could throw it when it was round your arm; this is truly an excellent trick. Various other devices may be used, but, as they are of no greater importance, we have omitted them.

343

PRINCIPLES OF DEFENCE AGAINST A DAGGER WITH
THE BARE HANDS.

It sometimes happens that an unarmed man is attacked by
an enemy with a dagger, who rushes at him with intent to mur-
der him, when the assaulted man has no refuge to fly to and is
in certain danger of being hit and killed. Since we desire to
show how in such a case it is possible to defend and attack the
enemy, we shall explain some methods, leaving others to be explained by others at other times. In order that you may be
better persuaded and convinced by our instruction, you must consider two principles; first that the man who seizes his dagger
to attack another man, seeing that his opponent has nothing
with which to defend himself, at once runs forward to hit in
the first place that occurs to him, only fearing least his opponent should escape, before he can hit him; therefore he uses
no skill, so that the man who is attacked can more easily defend. The second point is, that the dagger is not long enough
to reach your body, while you are bending forward and extending
your arms towards the adversary's hands and hilt, whether his
stroke is high or low. Nor can his arm be so strong of itself
as to force your two extended arms to yield; your two hands
are almost always united, except when you have brought your
body out of line, or twisted the adversary's arm, which takes
away his strength, as will be shown in the first encounter,
where we explain how one hand alone can defend and dash the
dagger from the adversary's hand. If the man with the dagger

tries to use his left hand, you should then seize that arm and turn it with the elbow over your shoulders; by giving it a wrench downwards you will not merely dislocate the arm, but even break it; or you can close with the adversary in order to throw him to the ground, or seize his dagger arm behind the elbow with your left hand and make him turn his back; in either case you will prevent his attacking with the dagger. In order to avoid undue length and the multiplication of examples, we shell include only the case of the adversary's attacking you, when he sees you have no defence at all.

Pl. 185.

This plate shows the adversary with his dagger drawn from the scabbard and his arm raised to strike, while you are waiting for him to attack. It is included to show the manner in which he has moved and holds his dagger; in the next plate the result will be shown; but afterwards we shall show the hit only, while the text will explain the position from which it has followed.

345

Pl. 186.

This disarm has arisen after the adversary has raised his dagger to strike, while you were waiting. He has aimed a blow downwards in continuation of his movement. You have raised your left arm, with your hand reversed and seized his arm as it fell close to the hilt of his dagger, giving the dagger a twist, as shown. Thus he could not prevent the dagger falling from his hand owing to the twist and the pain to his arm. He has bent his back in order to save his dagger; but as a result his position on his feet has been weakened,so that he is more likely to fall to the ground under the blow of your right arm, when you will finally wrest the dagger from his hand.

Pl. 187.

The next disarm, causing the dagger to fall from the adversary's hand, has happened in this way: he has driven the point of his dagger at your body from below; you were standing with your hands raised; you have placed your right hand on his blade, and your left hand under his dagger hand, lifting it up and pressing down the blade with your right hand, so that you have weakened his hand and easily forced the dagger out of his hand.

Pl. 188.

Here again the adversary has lost his dagger. He has driven
the point of his dagger straight forward with his hand in _tierce_;
with your hands close together, you have seized his hand and
hilt of his dagger, with your body bent low; pressing downwards
with your body and right hand on his hilt and lifting his arm
with your left hand, you have torn the dagger from his hand with
ease. The plate shows the position after it has left his hand.

Pl. 189.

Here is another disarm and hit. The adversary with the dagger has tried to hit you who are unarmed,thrusting the point at
your body from below upwards. You have placed your right hand
under the blade of the dagger, and your left hand over his dagger hand; by drawing his hand towards you, and pushing your
right hand forward, you have turned the point against him.
By resting your chest against the pommel of the dagger and throwing the whole of your weight on to it, you have driven the
point into the chest of the man who was holding it. His only
chance of safety was to drop the dagger to the ground, but he
should have done that when you began to turn his hand, for after it was turned, the point would have reached his body.

Let this suffice on the subject of the bare hands against
the dagger.

347

Pl. 190.

DEFENDING WITH THE SWORD ALONE AGAINST A PIKE.

The last plate illustrates a position not treated of by other writers. The point of the sword is held at right angles and directed to the ground in order to show the position of the body and sword in attacking a spontoon or half pike or other weapons, whether slightly longer or shorter matters not; it is equally unimportant whether the blade is longer or shorter,provided there are no barbs or other impediment. The left hand is in the greatest danger, but by proceeding in the correct manner you may easily protect it, that is by raising or lowering it more or less according to the line of the attack. You can also defend yourself against feints, disengagements, withdrawal or advancing of the pike, as well as against the simple thrust. You can defend very well against the cut too. You must advance without stopping for any reason, and although the arms are so unequal, by proceeding is the correct manner you will force the adversary to retire, or you will reach him more quickly and easily. We have omitted further details in order not to expose the whole secret, and in order to offer a subject to those who study this art or investigating the principles suited to this defence; by diligent practise you may with no great difficulty discover what is needed. The plate shows the position of the sword and body, and by no very long practise a keen intelligence will understand the advantage of the position and will learn how to use it. It is sufficient for us to have given the hint and

shown that with the sword alone you can attack and defend at a pike, perhaps more easily than another sword, as we have seen in actual practice many times and on different occasions in the presence of gentlemen and great princes.

Among the plates of the guards, movements and hits in this work there are some which are defective in the grasp of the weapons, the position of the hilts, the turning of the hands, and the posture of the feet and body. In reality these positions are free and unstrained, otherwise the movements would be too slow. Still we hope that the discourses will supply the defects and explain what is to be inferred from the plates. The author began to divide the work into chapters, but being prevented by various difficulties he has allowed it to be printed and to remain as it is, divided into two books and each book into three parts.

The printing of these discourses was completed on September the 25th, 1606, in the City of Copenhagen, the metropolis of the Kingdom of Denmark, in the house of Hendrich Walchirchen.

www.ingramcontent.com/pod-product-compliance
Lightning Source LLC
Chambersburg PA
CBHW081004140626
46546CB00019B/3186